Student Solutions Manual and Study Guide

for

Discrete Mathematics with Applications

Third Edition

Susanna S. Epp
DePaul University

with the assistance of

Tom Jenkyns
Brock University

BROOKS/COLE
CENGAGE Learning

Australia • Brazil • Japan • Korea • Mexico • Singapore • Spain • United Kingdom • United States

BROOKS/COLE
CENGAGE Learning™

For product information and technology assistance, contact us at
Cengage Learning Customer & Sales Support, 1-800-354-9706

For permission to use material from this text or product, submit all requests online at
www.cengage.com/permissions
Further permissions questions can be emailed to
permissionrequest@cengage.com

ISBN-13: 978-0-534-36028-3
ISBN-10: 0-534-36028-9

Brooks/Cole
10 Davis Drive
Belmont, CA 94002-3098
USA

Cengage Learning is a leading provider of customized learning solutions with office locations around the globe, including Singapore, the United Kingdom, Australia, Mexico, Brazil, and Japan. Locate your local office at: **www.cengage.com/global**

Cengage Learning products are represented in Canada by Nelson Education, Ltd.

To learn more about Brooks/Cole, visit **www.cengage.com/brookscole**

Purchase any of our products at your local college store or at our preferred online store **www.ichapters.com**

Printed in the United States of America
6 7 8 9 10 12 11 10

Table of Contents

Note to Students

This *Student Solutions Manual and Study Guide* for the third edition of *Discrete Mathematics with Applications* contains complete solutions to every third exercise in the text that is not fully answered in Appendix B. It also contains additional explanation, and review material for each chapter.

The specific topics developed in the book provide a foundation for virtually every other mathematics or computer science subject you might study in the future. Perhaps more important, however, is the book's recurring focus on the logical principles used in mathematical reasoning and the techniques of mathematical proof.

Why should it matter for you to learn these things? The main reason is that these principles and techniques are the foundation for all kinds of careful analyses, whether of mathematical statements, computer programs, legal documents, or other technical writing. A person who understands and knows how to develop basic mathematical proofs has learned to think in a highly disciplined way, is able to deduce correct consequences from a few basic principles, can build a logically connected chain of statements and appreciates the need for giving a valid reason for each statement in the chain, is able to move flexibly between abstract symbols and concrete objects, and can deal with multiple levels of abstraction. Mastery of these skills opens a host of interesting and rewarding possibilities in a person's life.

In studying the subject matter of this book, you are embarking on an exciting and challenging adventure. I wish you much success!

Acknowledgements

I am enormously indebted to the work of Tom Jenkyns, whose eagle eye, mathematical knowledge, and understanding of language made an invaluable contribution to this volume. I am also most grateful to my husband, Helmut Epp, who constructed all the diagrams and provided much support and wise counsel over many years.

Susanna S. Epp

Chapter 1: The Logic of Compound Statements

The ability to reason using the principles of logic is essential for solving problems in abstract mathematics and computer science and for understanding the reasoning used in mathematical proof and disproof. In this chapter the various rules used in logical reasoning are developed both symbolically and in the context of their somewhat limited but very important use in everyday language.

Exercise sets for Sections 1.1–1.3 and 2.1–2.4 contain sentences for you to negate, write the contrapositive for, and so forth. These are designed to help you learn to incorporate the rules of logic into your general reasoning processes. Chapters 1 and 2 also present the rudiments of symbolic logic as a foundation for a variety of upper-division courses. Symbolic logic is used in, among others, the study of digital logic circuits, relational databases, artificial intelligence, and program verification.

Section 1.1

9. $(n \vee k) \wedge \sim (n \wedge k)$

15. When you are filling out a truth table, a convenient way to remember the definitions of $\sim$ (not), $\wedge$ (and), and $\vee$ (or) is to think of them in words as follows.

(1) A *not* statement has opposite truth value from that of the statement. So to fill out a column of a truth table for the negation of a statement, you look at the column representing the truth values for the statement. In each row where the truth value for the statement is T, you place an F in the corresponding row in the column representing the truth value of the negation. In each row where the truth value for the statement is F, you place a T in the corresponding row in the column representing the truth value of the negation.

(2) The only time an *and* statement is true is when both components are true. Thus to fill out a column of a truth table where the word *and* links two component statements, just look at the columns with the truth values for the component statements. In any row where both columns have a T, you put a T in the same row in the column for the *and* statement. In every other row of this column, you put a F.

(3) The only time an *or* statement is false is when both components are false. So to fill out a column of a truth table where the word *or* links two component statements, you look at the columns with the truth values for the component statements. In any row where both columns have a F, you put a F in the same row in the column for the *or* statement. In every other row of this column, you put a T.

p	q	$p \wedge q$	$p \vee q$	$\sim (p \wedge q)$	$(p \wedge q) \vee \sim (p \vee q)$
T	T	T	T	F	T
T	F	F	T	T	T
F	T	F	T	T	T
F	F	F	F	T	T

18.

p	q	r	$\sim p$	$\sim r$	$\sim p \vee q$	$q \wedge \sim r$	$\sim (q \wedge \sim r)$	$p \vee (\sim p \vee q)$	$(p \vee (\sim p \vee q)) \wedge \sim (q \wedge \sim r)$
T	T	T	F	F	T	F	T	T	T
T	T	F	F	T	T	T	F	T	F
T	F	T	F	F	F	F	T	T	T
T	F	F	F	T	F	F	T	T	T
F	T	T	T	F	T	F	T	T	T
F	T	F	T	T	T	T	F	T	F
F	F	T	T	F	T	F	T	T	T
F	F	F	T	T	T	F	T	T	T

24.

p	q	r	$q \vee r$	$p \wedge q$	$p \wedge r$	$p \wedge (q \vee r)$	$(p \wedge q) \vee (p \wedge r)$
T	T	T	T	T	T	T	T
T	T	F	T	T	F	T	T
T	F	T	T	F	T	T	T
T	F	F	F	F	F	F	F
F	T	T	T	F	F	F	F
F	T	F	T	F	F	F	F
F	F	T	T	F	F	F	F
F	F	F	F	F	F	F	T

same truth values

The truth table shows that $p \wedge (q \vee r)$ and $(p \wedge q) \vee (p \wedge r)$ always have the same truth values. Therefore they are logically equivalent. This proves the distributive law for $\wedge$ over $\vee$.

30. Sam is not an orange belt or Kate is not a red belt.

33. The dollar is not at an all-time high or the stock market is not at a record low.

36. $-10 \geq x$ or $x \geq 2$

48. *Solution 1:* $p \wedge (\sim q \vee p)$ $\equiv$ $p \wedge (p \vee \sim q)$ commutative law for $\vee$
$\equiv$ p absorption law

   Solution 2: $p \wedge (\sim q \vee p)$ $\equiv$ $(p \wedge \sim q) \vee (p \wedge p)$ distributive law
$\equiv$ $(p \wedge \sim q) \vee p$ identity law for $\wedge$
$\equiv$ p by the steps of exercise 47.

51. $(p \wedge (\sim (\sim p \vee q))) \vee (p \wedge q)$ $\equiv$ $(p \wedge (\sim (\sim p) \wedge \sim q)) \vee (p \wedge q)$ De Morgan's law
$\equiv$ $(p \wedge (p \wedge \sim q)) \vee (p \wedge q)$ double negative law
$\equiv$ $((p \wedge p) \wedge \sim q)) \vee (p \wedge q)$ associative law for $\wedge$
$\equiv$ $(p \wedge \sim q)) \vee (p \wedge q)$ idempotent law for $\wedge$
$\equiv$ $p \wedge (\sim q \vee q)$ distributive law
$\equiv$ $p \wedge (q \vee \sim q)$ commutative law for $\vee$
$\equiv$ $p \wedge \mathbf{t}$ negation law for $\vee$
$\equiv$ p identity law for $\wedge$

54. The conditions are most easily symbolized as $p \vee (q \wedge \sim (r \wedge (s \wedge t)))$, but may also be written in a logically equivalent form.

Section 1.2

6. The only time an *if-then* ($\rightarrow$) statement is false is when the hypothesis is true and the conclusion is false. So to fill out a column of a truth table for an *if-then* statement, you look at the columns with the truth values for its hypothesis and its conclusion. In any row where the column with the truth values for the hypothesis has a T and where the column with the truth values for the conclusion has an F, you put a F in the same row in the column for the *if-then* statement. In every other row of this column, you put a T.

p	q	$\sim p$	$\sim p \wedge q$	$p \vee q$	$(p \vee q) \vee (\sim p \wedge q)$	$(p \vee q) \vee (\sim p \wedge q) \rightarrow q$
T	T	F	F	T	T	T
T	F	F	F	T	T	F
F	T	T	T	T	T	T
F	F	T	F	F	F	T

15.

p	q	r	q → r	p → q	p → (q → r)	(p → q) → r
T	T	T	T	T	T	T
T	T	F	F	T	F	F
T	F	T	T	F	T	T
T	F	F	T	F	T	T
F	T	T	T	T	T	T
F	T	F	F	T	T	F
F	F	T	T	T	T	F
F	F	F	T	T	T	F

different truth values

The truth table shows that $p \to (q \to r)$ and $(p \to q) \to r$ do not always have the same truth values. (They differ for the combinations of truth values for p, q, and r shown in rows 6, 7, and 8.) Therefore they are not logically equivalent.

18. *Part 1*: Let p represent "It walks like a duck," q represent "It talks like a duck," and r represent "It is a duck." The statement "If it walks like a duck and it talks like a duck, then it is a duck" has the form $p \land q \to r$. And the statement "Either it does not walk like a duck or it does not talk like a duck or it is a duck" has the form $\sim p \lor \sim q \lor r$.

p	q	r	~ p	~ q	p ∧ q	~ p ∨ ~ q	p ∧ q → r	(~ p ∨ ~ q) ∨ r
T	T	T	F	F	T	F	T	T
T	T	F	F	F	T	F	F	F
T	F	T	F	T	F	T	T	T
T	F	F	F	T	F	T	T	T
F	T	T	T	F	F	T	T	T
F	T	F	T	F	F	T	T	T
F	F	T	T	T	F	T	T	T
F	F	F	T	T	F	T	T	T

same truth values

The truth table shows that $p \land q \to r$ and $(\sim p \lor \sim q) \lor r$ always have the same truth values. Thus the following statements are logically equivalent: "If it walks like a duck and it talks like a duck, then it is a duck" and "Either it does not walk like a duck or it does not talk like a duck or it is a duck."

Part 2: The statement "If it does not walk like a duck and it does not talk like a duck then it is not a duck" has the form $\sim p \land \sim q \to \sim r$.

p	q	r	~ p	~ q	~ r	p ∧ q	~ p ∧ ~ q	p ∧ q → r	(~ p ∧ ~ q) → ~ r
T	T	T	F	F	F	T	F	T	T
T	T	F	F	F	T	T	F	F	T
T	F	T	F	T	F	F	F	T	T
T	F	F	F	T	T	F	F	T	T
F	T	T	T	F	F	F	F	T	T
F	T	F	T	F	T	F	F	T	T
F	F	T	T	T	F	F	T	T	F
F	F	F	T	T	T	F	T	T	T

different truth values

The truth table shows that $p \land q \to r$ and $(\sim p \land \sim q) \to \sim r$ do not always have the same truth values. (They differ for the combinations of truth values of p, q, and r shown in rows 2 and 7.) Thus they are not logically equivalent, and so the statement "If it walks like a duck and it talks like a duck, then it is a duck" is not logically equivalent to the statement "If it does not walk like a duck and it does not talk like a duck then it is not a duck." In addition, because of the logical equivalence shown in Part 1, we can also conclude that the following two statements are not logically equivalent: "Either it does not walk like a duck or it does not talk like a duck or it is a duck" and "If it does not walk like a duck and it does not talk like a duck then it is not a duck."

21. By the truth table for $\rightarrow$, $p \rightarrow q$ is false if, and only if, p is true and q is false. Under these circumstances, (b) $p \lor q$ is true and (c) $q \rightarrow p$ is also true.

    The truth table shows that $p \rightarrow q$ and $\sim p \rightarrow \sim q$ have different truth values in rows 2 and 3, so they are not logically equivalent. Thus a conditional statement is not logically equivalent to its inverse.

27.

p	q	$\sim p$	$\sim q$	$q \rightarrow p$	$\sim p \rightarrow \sim q$
T	T	F	F	T	T
T	F	F	T	T	T
F	T	T	F	F	F
F	F	T	T	T	T

same truth values

The truth table shows that $q \rightarrow p$ and $\sim p \rightarrow \sim q$ always have the same truth values, so they are logically equivalent. Thus the converse and inverse of a conditional statement are logically equivalent to each other.

30. The corresponding tautology is $p \land (q \lor r) \leftrightarrow (p \land q) \lor (p \land r)$

p	q	r	$q \lor r$	$p \land q$	$p \land r$	$p \land (q \lor r)$	$(p \land q) \lor (p \land r)$	$p \land (q \lor r) \leftrightarrow$ $(p \land q) \lor (p \land r)$
T	T	T	T	T	T	T	T	T
T	T	F	T	T	F	T	T	T
T	F	T	T	F	T	T	T	T
T	F	F	F	F	F	F	F	T
F	T	T	T	F	F	F	F	T
F	T	F	T	F	F	F	F	T
F	F	T	T	F	F	F	F	T
F	F	F	F	F	F	F	F	T

all T's

The truth table shows that $p \land (q \lor r) \leftrightarrow (p \land q) \lor (p \land r)$ is always true. Hence it is a tautology.

33. If Sam is not an expert sailor, then he will not be allowed on Signe's racing boat.

    If Sam is allowed on Signe's racing boat, then he is an expert sailor.

36. If it doesn't rain, then Ann will go.

39. a. $p \lor \sim q \rightarrow r \lor q$ $\equiv$ $\sim (p \lor \sim q) \lor (r \lor q)$ *[an acceptable answer]*
    $\equiv$ $(\sim p \land \sim (\sim q)) \lor (r \lor q)$ by De Morgan's law
    [another acceptable answer]
    $\equiv$ $(\sim p \land q) \lor (r \lor q)$ by the double negative law
    [another acceptable answer]

b. $p \vee \sim q \to r \vee q$ $\equiv$ $(\sim p \wedge q) \vee (r \vee q)$ by part (a)
            $\equiv$ $\sim (\sim (\sim p \wedge q) \wedge \sim (r \vee q))$ by De Morgan's law
            $\equiv$ $\sim (\sim (\sim p \wedge q) \wedge (\sim r \wedge \sim q))$ by De Morgan's law

The steps in the answer to part (b) would also be acceptable answers for part (a).

42. Yes. As in exercises 29–32, the following logical equivalences can be used to rewrite any statement form in a logically equivalent way using only $\sim$ and $\wedge$:

$$p \to q \equiv \sim p \vee q \qquad\qquad p \leftrightarrow q \equiv (\sim p \vee q) \wedge (\sim q \vee p)$$
$$p \vee q \equiv \sim (\sim p \wedge \sim q) \qquad\qquad \sim (\sim p) \equiv p$$

The logical equivalence $p \wedge q \equiv \sim (\sim p \vee \sim q)$ can then be used to rewrite any statement form in a logically equivalent way using only $\sim$ and $\vee$.

48. If this computer program produces error messages during translation, then it is not correct.

If this computer program is correct, then it does not produce error messages during translation.

Section 1.3

9.

						premises			conclusion	
p	q	r	$\sim q$	$\sim r$	$p \wedge q$	$p \wedge q \to \sim r$	$p \vee \sim q$	$\sim q \to p$	$\sim r$	
T	T	T	F	F	T	T	T	T	F	critical row
T	T	F	F	T	T	F	T	T	T	
T	F	T	T	F	F	T	T	T	F	critical row
T	F	F	T	T	F	T	T	T	T	
F	T	T	F	F	F	T	F	T	F	
F	T	F	F	T	F	T	F	T	T	
F	F	T	T	F	F	T	T	F	F	
F	F	F	T	T	F	T	T	F	T	

Rows 1 and 3 of the truth table show that it is possible for an argument of this form to have true premises and a false conclusion. Hence the argument form is invalid.

12.

		premises		conclusion
p	q	$p \to q$	$\sim q$	$\sim p$
T	T	T	F	F
T	F	F	T	F
F	T	T	F	T
F	F	T	T	T ← critical row

The truth table shows that in the only situation (represented by row 4) in which both premises are true, the conclusion is also true. Therefore, modus tollens is valid.

15.

		premise	conclusion
p	q	q	$p \vee q$
T	T	T	T ← critical row
T	F	F	T
F	T	T	T ← critical row
F	F	F	F

The truth table shows that in the two situations (represented by rows 1 and 3) in which the premise is true, the conclusion is also true. Therefore, the second version of generalization is valid.

21.

			premises			conclusion
p	q	r	$p \vee q$	$p \rightarrow r$	$q \rightarrow r$	r
T	T	T	T	T	T	T ← ——— critical row
T	T	F	T	F	F	F
T	F	T	T	T	T	T ← ——— critical row
T	F	F	T	F	T	F
F	T	T	T	T	T	T ← ——— critical row
F	T	F	T	T	F	F
F	F	T	F	T	T	T
F	F	F	F	T	T	F

The truth table shows that in the three situations (represented by rows 1, 3, 5) in which all three premises are true, the conclusion is also true. Therefore, proof by division into cases is valid.

30. form: $p \rightarrow q$ invalid, converse error
$$q$$
$$\therefore \quad p$$

33. An invalid argument with a true conclusion can have premises that are either true or false. In the following example the first premise is true for either one of two reasons: its hypothesis is false and its conclusion is true.

If the square of every real number is positive, then some real numbers are positive.

Some real numbers are positive.

Therefore, the square of every real number is positive.

42. (1) $q \rightarrow r$ premise b
$\sim r$ premise d
$\therefore \quad \sim q$ by modus tollens

(2) $p \vee q$ premise a
$\sim q$ by (1)
$\therefore \quad p$ by elimination

(3) $\sim q \rightarrow u \wedge s$ premise e
$\sim q$ by (1)
$\therefore \quad u \wedge s$ by modus ponens

(4) $u \wedge s$ by (3)
$\therefore \quad s$ by specialization

(5) p by (2)
s by (4)
$\therefore \quad p \wedge s$ by conjunction

(6) $p \wedge s \rightarrow t$ premise c
$p \wedge s$ by (5)
$\therefore \quad t$ by modus ponens

Section 1.4

6. The input/output table is as follows:

Input		Output
P	Q	R
1	1	0
1	0	1
0	1	0
0	0	0

12. $(P \lor Q) \lor \sim (Q \land R)$

15.

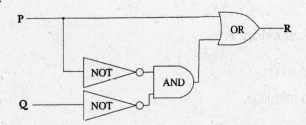

21. *a.* $(P \land Q \land \sim R) \lor (\sim P \land Q \land R) \lor (\sim P \land Q \land \sim R)$

b. One circuit having the given input/output table is the following:

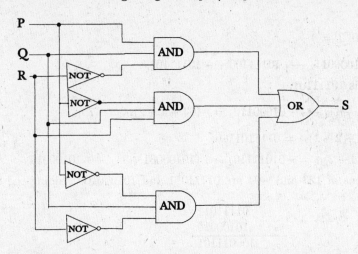

27. The Boolean expression for circuit (a) is $\sim P \land (\sim (\sim P \land Q))$ and for circuit (b) it is $\sim (P \lor Q)$. We must show that if these expressions are regarded as statement forms, then they are logically equivalent. But

$$
\begin{aligned}
\sim P \land (\sim (\sim P \land Q)) &\equiv \sim P \land (\sim (\sim P) \lor \sim Q) && \text{by De Morgan's law} \\
&\equiv \sim P \land (P \lor \sim Q) && \text{by the double negative law} \\
&\equiv (\sim P \land P) \lor (\sim P \land \sim Q) && \text{by the distributive law} \\
&\equiv \mathbf{c} \lor (\sim P \land \sim Q) && \text{by the negation law for } \land \\
&\equiv \sim P \land \sim Q && \text{by the identity law for } \lor \\
&\equiv \sim (P \lor Q) && \text{by De Morgan's law.}
\end{aligned}
$$

33. *a.*

$$
\begin{aligned}
(P \mid Q) \mid (P \mid Q) &\equiv \ \sim [(P \mid Q) \wedge (P \mid Q)] &&\text{by definition of } \mid \\
&\equiv \ \sim (P \mid Q) &&\text{by the idempotent law for } \wedge \\
&\equiv \ \sim [\sim (P \wedge Q)] &&\text{by definition of } \mid \\
&\equiv \ P \wedge Q &&\text{by the double negative law.}
\end{aligned}
$$

1. *b.*

$$
\begin{aligned}
P \wedge (\sim Q \vee R) &\equiv (P \mid (\sim Q \vee R)) \mid (P \mid (\sim Q \vee R)) \\
&\qquad\qquad\qquad\qquad\qquad\qquad\qquad \text{by part (a)} \\
&\equiv (P \mid [(\sim Q \mid \sim Q) \mid (R \mid R)]) \mid (P \mid [(\sim Q \mid \sim Q) \mid (R \mid R)]) \\
&\qquad\qquad\qquad\qquad\qquad\qquad\qquad \text{by Example 1.4.7(b)} \\
&\equiv (P \mid [((Q \mid Q) \mid (Q \mid Q)) \mid (R \mid R)]) \mid \\
&\qquad (P \mid [((Q \mid Q) \mid (Q \mid Q)) \mid (R \mid R)]) \\
&\qquad\qquad\qquad\qquad\qquad\qquad\qquad \text{by Example 1.4.7(a)}
\end{aligned}
$$

Section 1.5

3. $287 = 256 + 16 + 8 + 4 + 2 + 1 = 100011111_2$

6. $1424 = 1024 + 256 + 128 + 16 = 10110010000_2$

9. $110110_2 = 32 + 16 + 4 + 2 = 54_{10}$

18.
$$
\begin{array}{r}
11010_2 \\
- \quad 1101_2 \\
\hline
1101_2
\end{array}
$$

21. *b.* $S = 0, T = 1$ *c.* $S = 0, T = 0$

24. $67_{10} = (64 + 2 + 1)_{10} = 01000011_2 \longrightarrow 10111100 \longrightarrow 10111101.$

So the two's complement is 10111101.

30. $10111010 \longrightarrow -(01000101 + 1)_2 \longrightarrow -01000110_2 = -(64 + 4 + 2)_{10} = -70_{10}$

36. $123_{10} = (64 + 32 + 16 + 8 + 2 + 1)_{10} = 01111011_2$

$-94_{10} = -(64 + 16 + 8 + 4 + 2)_{10} = -01011110_2 \longrightarrow (10100001 + 1)_2 \longrightarrow 10100010$

So the 8-bit representations of 123 and -94 are 01111011 and 10100010. Adding the 8-bit representations gives

$$
\begin{array}{r}
01111011 \\
+ \quad 10100010 \\
\hline
100011101
\end{array}
$$

Truncating the 1 in the 2^8th position gives 00011101. Since the leading bit of this number is a 0, the answer is positive. Converting back to decimal form gives

$$
00011101 \longrightarrow 11101_2 = (16 + 8 + 4 + 1)_{10} = 29_{10}.
$$

So the answer is 29.

39. $E0D_{16} = 14 \cdot 16^2 + 0 + 13 = 3597_{10}$

42. $B53DF8_{16} = 1011\ 0101\ 0011\ 1101\ 1111\ 1000_2$

45. $1011\ 0111\ 1100\ 0101_2 = B7C5_{16}$

General Review Guide: Chapter 1

Compound Statements

- What is a statement? *(p. 2)*
- If p and q are statements, how do you symbolize "p but q" and "neither p nor q"? *(p. 3)*
- What does the notation $a \leq x < b$ mean? *(p. 4)*
- What is the conjunction of statements p and q? *(p. 5)*
- What is the disjunction of statements p and q? *(p. 6)*
- What are the truth table definitions for $\sim p$, $p \wedge q$, $p \vee q$, $p \rightarrow q$, and $p \leftrightarrow q$? *(pp. 5, 6, 18, 24)*
- How do you construct a truth table for a general compound statement? *(p. 7)*
- What is exclusive or? *(p. 7)*
- What is a tautology, and what is a contradiction? *(p. 13)*
- What is a conditional statement? *(p. 18)*
- Given a conditional statement, what is its hypothesis (antecedent)? conclusion (consequent)? *(p. 18)*
- What is a biconditional statement? *(p. 24)*
- What is the order of operations for the logical operators? *(p. 24)*

Logical Equivalence

- What does it mean for two statement forms to be logically equivalent? *(p. 8)*
- How do you test to see whether two statement forms are logically equivalent? *(p. 9)*
- How do you annotate a truth table to explain how it shows that two statement forms are or are not logically equivalent? *(p. 9)*
- What is the double negative property? *(p. 9)*
- What are De Morgan's laws? *(p. 10)*
- How is Theorem 1.1.1 used to show that two statement forms are logically equivalent? *(p. 14)*
- What are negations for the following forms of statements? *(pp. 10, 11, 20)*
    - $p \wedge q$
    - $p \vee q$
    - $p \rightarrow q$ (if p then q)

Converse, Inverse, Contrapositive

- What is the contrapositive of a statement of the form "If p then q"? *(p. 21)*
- What are the converse and inverse of a statement of the form "If p then q"? *(p. 22)*
- Can you express converses, inverses, and contrapositives of conditional statements in ordinary English? *(p. 21-22)*
- If a conditional statement is true, can its converse also be true? *(p. 22)*
- Given a conditional statement and its contrapositive, converse, and inverse, which of these are logically equivalent and which are not? *(p. 23)*

Necessary and Sufficient Conditions, Only If

- What does it mean to say that something is true only if something else is true? *(p. 23)*
- How are statements about only-if statements translated into if-then form.? *(p. 23)*
- What does it mean to say that something is a necessary condition for something else? *(p. 25)*
- What does it mean to say that something is a sufficient condition for something else? *(p. 25)*

- How are statements about necessary and sufficient conditions translated into if-then form.? *(pp. 25-26)*

Validity and Invalidity

- How do you identify the logical form of an argument? *(p. 2)*
- What does it mean for a form of argument to be valid? *(p. 29)*
- How do you test to see whether a given form of argument is valid? *(p. 30)*
- How do you annotate a truth table to explain how it shows that an argument is or is not valid? *(pp. 30-31)*
- What are modus ponens and modus tollens? *(pp. 31-32)*
- Can you give examples for and prove the validity of the following forms of argument? *(pp. 33-35)*

$$\begin{array}{c} p \\ \therefore \ p \lor q \end{array} \quad \text{and} \quad \begin{array}{c} q \\ \therefore \ p \lor q \end{array}$$

$$\begin{array}{c} p \land q \\ \therefore \ p \end{array} \quad \text{and} \quad \begin{array}{c} p \land q \\ \therefore \ q \end{array}$$

$$\begin{array}{c} p \lor q \\ \sim q \\ \therefore \ p \end{array} \quad \text{and} \quad \begin{array}{c} p \lor q \\ \sim p \\ \therefore \ q \end{array}$$

$$\begin{array}{c} p \to q \\ q \to r \\ \therefore \ p \to r \end{array}$$

$$\begin{array}{c} p \lor q \\ p \to r \\ q \to r \\ \therefore \ r \end{array}$$

- What are converse error and inverse error? *(p. 37)*
- Can a valid argument have a false conclusion? *(p. 38)*
- Can an invalid argument have a true conclusion? *(p. 38)*
- Which of modus ponens, modus tollens, converse error, and inverse error are valid and which are invalid? *(pp. 31, 32, 37, 38)*
- What is the contradiction rule? *(p. 39)*
- How do you use valid forms of argument to solve puzzles such as those of Raymond Smullyan about knights and knaves? *(p. 40)*

Digital Logic Circuits and Boolean Expressions

- Given a digital logic circuit, how do you
  - find the output for a given set of input signals *(p. 47)*
  - construct an input/output table *(p. 47)*
  - find the corresponding Boolean expression? *(p. 48)*
- What is a recognizer? *(p. 49)*
- Given a Boolean expression, how do you draw the corresponding digital logic circuit? *(p. 49)*
- Given an input/output table, how do you draw the corresponding digital logic circuit? *(p. 51)*
- What is disjunctive normal form? *(p. 52)*
- What does it mean for two circuits to be equivalent? *(p. 53)*
- What are NAND and NOR gates? *(p. 54)*
- What are Sheffer strokes and Peirce arrows? *(p. 54)*

Binary and Hexadecimal Notation

- How do you transform positive integers from decimal to binary notation and the reverse? *(p. 59)*
- How do you add and subtract integers using binary notation? *(p. 60)*
- What is a half-adder? *(p. 61)*
- What is a full-adder? *(p. 62)*
- What is the 8-bit two's complement of an integer in binary notation? *(p. 63)*
- How do you find the 8-bit two's complement of a positive integer a that is at most 255? *(p. 64)*
- How do you find the decimal representation of the integer with a given 8-bit two's complement? *(p. 65)*
- How are negative integers represented using two's complements? *(p. 66)*
- How is computer addition with negative integers performed? *(pp. 66-70)*
- How do you transform positive integers from hexadecimal to decimal notation? *(p. 71)*
- How do you transform positive integers from binary to hexadecimal notation and the reverse? *(p. 72)*
- What is octal notation? *(p. 74)*

Test Your Understanding: Chapter 1

Test yourself by filling in the blanks.

1. A statement is ____.

2. If p and q are statements, the statement "p but q" is symbolized ____.

3. If p and q are statements, the statement "neither p nor q" is symbolized ____.

4. An *and* statement is true if, and only if, both components are ____.

5. An *or* statement is false if, and only if, both components are ____.

6. An *if-then* statement is false if, and only if, its hypothesis is ____ and its conclusion is ____.

7. A statement of the form $p \leftrightarrow q$ is true if, and only if, ____.

8. If a logical expression includes the symbols $\sim$, $\wedge$ or $\vee$, and $\rightarrow$ or $\leftrightarrow$ and the expression does not include parentheses, then the first operation to be performed is ____, the second is ____, and the third is ____. To indicate the order of operations for an expression that includes both $\wedge$ and $\vee$ or both $\rightarrow$ and $\leftrightarrow$, it is frequently necessary to add ____.

9. A tautology is ____ for every substitution of statements for the statement variables.

10. A contradiction is ____ for every substitution of statements for the statement variables.

11. Two statement forms are logically equivalent if, and only if, their truth values are ____ for every substitution of statements for the statement variables.

12. *Less formal version*: Two statement forms are logically equivalent if, and only if, they always have ____.

13. Two statement forms are not logically equivalent if, and only if, ____.

14. De Morgan's laws say that ____ and ____.

15. The negation of $p \rightarrow q$ is ____.

16. The contrapositive of "if p then q" is ____.

17. The converse of "if p then q" is ____.

18. The inverse of "if p then q" is ____.

19. A conditional statement and its contrapositive are ____.

20. A conditional statement and its converse are not ____.

21. If r and s are statements, r only if s can be expressed in if-then form as ____ or as ____.

22. If t and u are statements, t is a sufficient condition for u can be expressed in if-then form as ____.

23. If v and w are statements, v is a necessary condition for w can be expressed in if-then form as ____ or as ____.

24. A form of argument is valid if, and only if, for every substitution of statements for the statement variables, if ____ of the premises are ____, then the conclusion is ____.

25. *Less formal version*: A form of argument is valid if, and only if, in all cases where the premises are____, the conclusion is also ____.

26. A form of argument is invalid if, and only if, there is a substitution of statements for the statement variables that makes the premises ____ and the conclusion ____.

27. *Less formal version*: A form of argument is invalid if, and only if, it is possible for all the premises to be ____ and the conclusion ____.

28. Modus ponens is an argument of the form ____, and modus tollens is an argument of the form ____.

29. Converse error is an argument of the form ____, and inverse error is an argument of the form ____.

30. Insert the words "valid " or "invalid" as appropriate: modus ponens is ____; modus tollens is ____; converse error is ____; inverse error is ____.

31. The contradiction rule is an argument of the form ____.

32. The input/output table for a digital logic circuit is a table that shows ____.

33. The Boolean expression that corresponds to a digital logic circuit is ____.

34. A recognizer is a digital logic circuit that ____.

35. Two digital logic circuits are equivalent if, and only if, ____.

36. A NAND-gate is constructed by placing a ____ gate immediately following an ____ gate.

37. A NOR-gate is constructed by placing a ____ gate immediately following an ____ gate.

38. To represent a nonnegative integer in binary notation means to write it as a sum of products of the form ____, where ____.

39. To add integers in binary notation, you use the facts that $1_2 + 1_2 =$ ____ and $1_2 + 1_2 + 1_2 =$ ____.

40. To subtract integers in binary notation, you use the facts that $10_2 - 1_2 =$ ____ and $11_2 - 1_2 =$ ____.

41. A half-adder is a digital logic circuit that ____, and a full-adder is a digital logic circuit that ____.

42. The 8-bit two's complement of a positive integer a is ____.

43. To find the 8-bit two's complement of a positive integer a that is at most 255, you ____, ____, and ____.

44. If a is an integer with $-128 \le a \le 127$, the 8-bit representation of a is ____ if $a \ge 0$ and is ____ if $a < 0$.

45. To add two integers in the range -128 through 127 whose sum is also in the range -128 through 127, you ____, ____, ____, and ____.

46. To represent a nonnegative integer in hexadecimal notation means to write it as a sum of products of the form ____, where ____.

47. To convert a nonnegative integer from hexadecimal to binary notation, you ____ and ____.

Answers

1. a sentence that is true or false but not both
2. $p \wedge q$
3. $\sim p \wedge \sim q$
4. true
5. false
6. true, false
7. both p and q are true or both p and q are false
8. $\sim$; $\wedge$ or $\vee$; $\rightarrow$ or $\leftrightarrow$; parentheses
9. true
10. false
11. identical
12. the same truth values
13. there exist statements with the property that when the statements are substituted for the statement variables, one of the resulting statements is true and the other is false
14. $\sim (p \wedge q) \equiv \sim p \vee \sim q$; $\sim (p \vee q) \equiv \sim p \wedge \sim q$
15. $p \vee \sim q$
16. if $\sim q$ then $\sim p$
17. if q then p
18. if $\sim p$ then $\sim q$
19. logically equivalent
20. logically equivalent
21. if $\sim s$ then $\sim r$; if r then s
22. if t then u
23. if $\sim v$ then $\sim w$; if w then v
24. all; true; true
25. true; true
26. true; false
27. true; false

28.
$$p \to q \qquad\qquad p \to q$$
$$p \qquad\qquad\qquad \sim q$$
$$\therefore \ q \qquad\qquad \therefore \ \sim p$$

29.
$$p \to q \qquad\qquad p \to q$$
$$q \qquad\qquad\qquad \sim p$$
$$\therefore \ p \qquad\qquad \therefore \ \sim q$$

30. valid; valid; invalid; invalid

31.
$$p \to \mathbf{c}, \text{ where } \mathbf{c} \text{ is a contradiction}$$
$$\therefore \ \sim p$$

32. shows the output signal(s) that correspond to all possible combinations of input signals to the circuit

33. a logical expression that represents the input signals symbolically and indicates the successive actions of the logic gates on the input signals

34. outputs a 1 for exactly one particular combination of input signals and outputs 0's for all other combinations

35. they have the same input/output table

36. NOT; AND

37. NOT; OR

38. $d \cdot 2^n$; $d = 0$ or $d = 1$, and n is a nonnegative integer

39. 10_2; 11_2

40. 1_2; 10_2

41. outputs the sum of any two binary digits;
outputs the sum of any three binary digits

42. $2^8 - a$

43. write the 8-bit binary representation of a;
flip the bits
add 1 in binary notation

44. the 8-bit binary representation of a
the 8-bit binary representation of $2^8 - a$

45. convert both integers to their 8-bit binary representations
add the results using binary notation
truncate any leading 1
convert back to decimal form

46. $d \cdot 16^n$; $d = 0, 1, 2, \ldots 9, A, B, C, D, E, F$, and n is a nonnegative integer

47. write each hexadecimal digit in fixed 4-bit binary notation
juxtapose the results

Chapter 2: The Logic of Quantified Statements

Ability to use the logic of quantified statements correctly is necessary for doing mathematics because mathematics is, in a very broad sense, about quantity. The main purpose of this chapter is to familiarize you with the language of universal and existential statements. The various facts about quantified statements developed in this chapter are used extensively in Chapter 3 and are referred to throughout the rest of the book. Experience with the formalism of quantification is especially useful to students planning to study LISP or Prolog, program verification, or relational databases.

Section 2.1

6. *a.* When $m = 25$ and $n = 10$, the statement "m is a factor of n^2" is true because $n^2 = 100$ and $100 = 4 \cdot 25$. But the statement "m is a factor of n" is false because 10 is not a product of 25 times any integer. Thus the hypothesis is true and the conclusion is false, so the statement as a whole is false.

   b. $R(m, n)$ is also false when $m = 8$ and $n = 4$ because 8 is a factor of $4^2 = 16$, but 8 is not a factor of 4.

   c. When $m = 5$ and $n = 10$, both statements "m is a factor of n^2" and "m is a factor of n" are true because $n = 10 = 5 \cdot 2 = m \cdot 2$ and $n^2 = 100 = 5 \cdot 20 = m \cdot 20$. Thus both the hypothesis and conclusion of $R(m, n)$ are true, and so the statement as a whole is true.

   d. Here are examples of two kinds of correct answers:

   (1) Let $m = 2$ and $n = 6$. Then both statements "m is a factor of n^2" and "m is a factor of n" are true because $n = 6 = 2 \cdot 3 = m \cdot 3$ and $n^2 = 36 = 2 \cdot 18 = m \cdot 18$. Thus both the hypothesis and conclusion of $R(m, n)$ are true, and so the statement as a whole is true.

   (2) Let $m = 6$ and $n = 2$. Then both statements "m is a factor of n^2" and "m is a factor of n" are false because $n = 2 \neq 6 \cdot k$, for any integer k, and $n^2 = 4 \neq 6 \cdot j$, for any integer j. Thus both the hypothesis and conclusion of $R(m, n)$ are false, and so the statement as a whole is true.

12. *Counterexample*: Let $x = 1$ and $y = 1$, and note that $\sqrt{1+1} = \sqrt{2}$, $\sqrt{1} + \sqrt{1} = 1 + 1 = 2$, and $2 \neq \sqrt{2}$. (This is one counterexample among many. Any real numbers x and y with $xy \neq 0$ will produce a counterexample.)

15. *a. Some acceptable answers:* All squares are rectangles. If a figure is a square then that figure is a rectangle. Every square is a rectangle. All figures that are squares are rectangles. Any figure that is a square is a rectangle.

   b. Some acceptable answers: There is a set with sixteen subsets. Some set has sixteen subsets. Some sets have sixteen subsets. There is at least one set that has sixteen subsets.

18. *c.* $\forall s$, if $C(s)$ then $\sim E(s)$.

   d. $\exists x$ such that $C(s) \wedge M(s)$.

21. *b.* $\forall x$, if x is a valid argument with true premises, then x has a true conclusion.

   Or: $\forall$ arguments x, if x is valid and x has true premises then x has a true conclusion.

   Or: $\forall$ valid arguments x, if x has true premises then x has a true conclusion.

   d. $\forall$ integers m and n, if m and n are odd then mn is odd.

24. *a.* $\forall x$, if x is an integer then x is rational, but $\exists x$ such that x is rational and x is not an integer.

27. *a.* This statement translates as "There is a geometric figure that is both a rectangle and a square." This is true. As an example take any square; it is a rectangle whose sides all have the same length.

b. This statement translates as "There is a geometric figure that is a rectangle but is not a square." This is true. Any rectangle whose sides are not all of the same length is a rectangle that is not a square. For example, one pair of parallel sides could be twice as long as the other pair of parallel sides.

c. This statement translates as "Every square is a rectangle." This is true. A square is a rectangle satisfying the additional condition that all its sides have the same length.

28. *a.* This statement translates as "There is a prime number that is not odd." This is true. The number 2 is prime and it is not odd.

30. *b.* This statement translates as "For all real numbers x, if $x > 2$ then $x^2 > 4$," which is true.

d. This statement translates as "For all real numbers x, $x^2 > 4$ if, and only if, $|x| > 2$." This is true because $x^2 > 4$ if, and only if, $x > 2$ or $x < -2$, and $|x| > 2$ means that either $x > 2$ or $x < -2$.

Section 2.2

3. *b.* ∃ a computer C such that C does not have a CPU.

d. ∀ bands b, b has won fewer than 10 Grammy awards.

6. *b.* *Formal negation:* ∀ real numbers x, x is not rational.

Some acceptable informal negations: No real numbers are rational. All real numbers are irrational.

12. The proposed negation is not correct. *Correct negation:* There are an irrational number x and a rational number y such that xy is rational. Or: There are an irrational number and a rational number whose product is rational.

15. *b.* True   *d.* True

e. False: $x = 36$ is a counterexample because the ones digit of x is 6 and the tens digit is neither 1 nor 2.

21. ∃ an integer n such that n is prime and both n is not odd and $n \neq 2$.

Or: ∃ an integer n such that n is prime and n is neither odd nor equal to 2.

30. *Contrapositive:* ∀ integers d, if $d \neq 3$ then $6/d$ is not an integer.

Converse: ∀ integers d, if $d = 3$ then $6/d$ is an integer.

Inverse: ∀ integers d, if $6/d$ is not an integer, then $d \neq 3$.

36. *Contrapositive:* If an integer is not odd, then its square is not odd.

Converse: If an integer is odd, then its square is odd.

Inverse: If the square of an integer is not odd, then the integer is not odd.

39. If an integer is divisible by 8, then it is divisible by 4.

45. There is a function that is a polynomial but does not have a real root.

Section 2.3

3. *c.* Let $y = \frac{4}{3}$. Then $xy = \left(\frac{3}{4}\right)\left(\frac{4}{3}\right) = 1$.

6. True.

Given $x =$	Choose $y =$	Is y a circle above x, with a different color from x?
e	$a, b,$ or c	yes ✓
g	a or c	yes ✓
h	a or c	yes ✓
j	b	yes ✓

9. *b.* True. *Solution 1*: Let $x = 0$. Then for any real number r, $x + r = r + x = r$ because 0 is an identity for addition of real numbers. Thus, because every element in E is a real number, $\forall y \in E, x + y = y$.

   Solution 2: Let $x = 0$. Then $x + y = y$ is true for each individual element y of E:

Choose $x = 0$	Given $y =$	Is $x + y = y$?
	-2	yes: $0 + (-2) = -2$ ✓
	-1	yes: $0 + (-1) = -1$ ✓
	0	yes: $0 + 0 = 0$ ✓
	1	yes: $0 + 1 = 1$ ✓
	2	yes: $0 + 2 = 2$ ✓

12. *b.* Every student has seen Star Wars.

   e. There are two different students who have both seen the same movie.

   f. There are two different students, one of whom has seen all the movies that the other has seen.

15. *[For this exercise, there are other correct answers besides those shown.]*

   a. There is at least one book that everyone has read.

   b. *Negation*: Given any book, there is a person who has not read that book.

   Or: $\forall$ books b, $\exists$ a person p such that p has not read b.

   Or: There is no book that everyone has read.

27. *a.* The statement says that there are a circle and a square such that the circle is above the square and has the same color as the square. This is true. For example, circle a is above square j and a and j have the same color.

39. *b.* $\forall$ purposes under heaven p, $\exists$ a time t such that t is the time for p.

42. $\exists$ a real number $\varepsilon > 0$ such that $\forall$ real numbers $\delta > 0$, $\exists$ a real number x such that $a - \delta < x < a + \delta$ and either $L - \varepsilon \geq f(x)$ or $f(x) \geq L + \varepsilon$.

48. *a.* The statement says that given any object, we can find another object that has a different color. This is true because there are objects of all three colors. So, for example, if we are given a blue object, we can find another that is black or gray, and we can proceed similarly if we are given an object of either of the other two colors.

   b. $\forall x(\exists y(x \neq y \rightarrow \sim\text{SameColor}(x, y)))$

   c. $\exists x(\forall y(x \neq y \wedge \text{SameColor}(x, y)))$

54. These statements do not necessarily have the same truth values. For instance, let $D = \mathbf{R}$, the set of all real numbers, let $P(x)$ be "x is positive," and let $Q(x)$ be "x is negative." Then "$\exists x \in D, (P(x) \wedge Q(x))$" can be written "$\exists$ a real number x such that x is both positive and negative," which is false. On the other hand, "$(\exists x \in D, \; P(x)) \wedge (\exists x \in D, Q(x))$ can be written "$\exists$ a real number that is positive and $\exists$ a real number that is negative," which is true.

Section 2.4

6. This computer program is not correct.

12. invalid, inverse error

15. invalid, converse error

18. valid, universal modus tollens

24. Valid. The only drawing representing the truth of the premises also represents the truth of the conclusion.

27. Valid. The only drawing representing the truth of the premises also represents the truth of the conclusion.

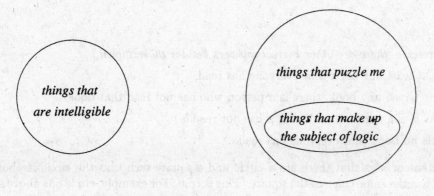

30. 3. If an object is black, then it is a square.

    2. (*contrapositive form*) If an object is a square, then it is above all the gray objects.

    4. If an object is above all the gray objects, then it is above all the triangles.

    1. If an object is above all the triangles, then it is above all the blue objects.

    ∴ If an object is black, then it is above all the blue objects.

36. The universal form of elimination (part a) says that the following form of argument is valid:

$$\forall x \text{ in } D, P(x) \vee Q(x). \qquad \leftarrow major\ premise$$
$$\sim Q(c) \text{ for a particular } c \text{ in } D. \qquad \leftarrow minor\ premise$$
$$\therefore \quad P(c)$$

Proof of Validity: Suppose the major and minor premises of the above argument form are both true. *[We must show the truth of the conclusion $P(c)$.]* By definition of truth value for a universal statement, $\forall x$ in D, $P(x) \vee Q(x)$ is true if, and only if, the statement "$P(x) \vee Q(x)$" is true for each individual element of D. So, by universal instantiation, it is true for the particular element c. Hence "$P(c) \vee Q(c)$" is true. And since the minor premise says that $\sim Q(c)$, it follows by the elimination rule that $P(c)$ is true. *[This is what was to be shown.]*

General Review Guide: Chapter 2

Quantified Statements

- What is a predicate? *(p. 76)*
- What is the truth set of a predicate? *(p. 77)*
- What is a universal statement, and what is required for such a statement to be true? *(p. 78)*
- What is the method of exhaustion? *(p. 79)*
- What is required for a universal statement to be false? *(p. 78)*
- What is an existential statement, and what is required for such a statement to be true? *(p. 80)*
- What is required for a existential statement to be false? *(p. 80)*
- What are some ways to translate quantified statements from formal to informal language? *(p. 80)*
- What are some ways to translate quantified statements from informal to formal language? *(pp. 81-82)*
- What is a universal conditional statement? *(p. 81)*
- What is an equivalent way to write a universal conditional statement? *(pp. 83)*
- What are equivalent ways to write existential statements? *(p. 83)*
- What does it mean for a statement to be quantified implicitly? *(p. 83)*
- What do the notations $\Rightarrow$ and $\Leftrightarrow$ mean? *(p. 84)*
- What is the relation among $\forall$, $\exists$, $\wedge$, and $\vee$? *(p. 91)*
- What does it mean for a universal statement to be vacuously true? *(p. 92)*
- What is the rule for interpreting a statement that contains both a universal and an existential quantifier? *(p. 99)*
- How are statements expressed in the computer programming language Prolog? *(p. 107)*

Negations: What are negations for the following forms of statements?

- $\forall x, Q(x)$ *(p. 88)*
- $\exists x$ such that $Q(x)$ *(p. 89)*
- $\forall x$, if $P(x)$ then $Q(x)$ *(p. 91)*
- $\forall x, \exists y$ such that $P(x,y)$ *(p. 103)*
- $\exists x$ such that $\forall y, P(x,y)$ *(p. 103)*

Variants of Conditional Statements

- What are the converse, inverse, and contrapositive of a statement of the form "$\forall x$, if $P(x)$ then $Q(x)$"? *(p. 93)*
- How are quantified statements involving necessary and sufficient conditions and the phrase only-if translated into if-then form? *(p. 95)*

Validity and Invalidity

- What is universal instantiation? *(p. 111)*
- What are the universal versions of modus ponens, modus tollens, converse error, and inverse error, and which of these forms of argument are valid and which are invalid? *(pp. 112, 114, 118)*
- How is universal modus ponens used in a proof? *(p. 113)*

- How can diagrams be used to test the validity of an argument with quantified statements? *(p. 115)*

Test Your Understanding: Chapter 2

Test yourself by filling in the blanks.

1. A predicate is ____.

2. The truth set of a predicate $P(x)$ with domain D is ____.

3. A statement of the form "$\forall x \in D, Q(x)$" is true if, and only if, ____.

4. A statement of the form "$\exists x \in D, Q(x)$" is true if, and only if, ____.

5. A universal conditional statement is a statement of the form ____.

6. A negation of a universal statement is an ____ statement.

7. A negation of an existential statement is a ____ statement.

8. A statement of the form "All A are B" can be written with a quantifier and a variable as ____.

9. A statement of the form "Some A are B" can be written with a quantifier and a variable as .

10. A statement of the form "No A are B" can be written with a quantifier and a variable as .

11. A negation for a statement of the form "$\forall x \in D, Q(x)$" is ____.

12. A negation for a statement of the form "$\exists x \in D$ such that $Q(x)$" is ____.

13. A negation for a statement of the form "$\forall x \in D$, if $P(x)$ then $Q(x)$" is ____.

14. For a statement of the form "$\forall x \in D, Q(x)$" to be vacuously true means that ____.

15. Given a statement of the form "$\forall x$, if $P(x)$ then $Q(x)$," the contrapositive is ____, the converse is ____, and the inverse is ____.

16. If you want to establish the truth of a statement of the form "$\forall x \in D, \exists y \in E$ such that $P(x,y)$," your challenge is to allow someone else to pick ____, and then you must find ____ for which $P(x,y)$ ____.

17. If you want to establish the truth of a statement of the form "$\exists x \in D$ such that $\forall y \in E, P(x,y)$," your job is to find ____ with the property that no matter what ____, $P(x,y)$ will be ____.

18. A negation for a statement of the form "$\forall x \in D, \exists y \in E$ such that $P(x,y)$" is ____.

19. A negation for a statement of the form "$\exists x \in D$ such that $\forall y \in E, P(x,y)$" is ____.

20. The rule of universal instantiation says that ____.

21. Universal modus ponens is an argument of the form ____, and universal modus tollens is an argument of the form ____.

22. To use a diagram to represent a statement of the form "All A are B," you ____.

23. To use a diagram to represent a statement of the form "Some A are B," you ____.

24. To use a diagram to represent a statement of the form "No A are B," you ____.

Answers

1. a sentence that contains a finite number of variables and becomes a statement when specific values are substituted for the variables

2. the set of all x in D such that $P(x)$ is true

3. $Q(x)$ is true for each individual x in D

4. there is at least one x in D for which $Q(x)$ is true

5. $\forall x$, if $P(x)$ then $Q(x)$, where $P(x)$ and $Q(x)$ are predicates

6. existential

7. universal

8. $\forall x$, if x is an A then x is a B

9. $\exists x$ such that x is an A and x is a B

10. $\forall x$, if x is an A then x is not a B (Or: $\forall x$, if x is an B then x is not a A)

11. $\exists x \in D$ such that $\sim Q(x)$

12. $\forall x \in D$, $\sim Q(x)$

13. $\exists x \in D$ such that $P(x)$ and $\sim Q(x)$

14. there are no elements in D

15. $\forall x$, if $\sim Q(x)$ then $\sim P(x)$;
    $\forall x$, if $Q(x)$ then $P(x)$
    $\forall x$, if $\sim P(x)$ then $\sim Q(x)$

16. whatever element x in D they wish; an element y in E;   is true

17. an element x in D; element y in E anyone might choose; true

18. $\exists x \in D$ such that $\forall y \in E$, $\sim P(x, y)$

19. $\forall x \in D$, $\exists y \in E$ such that $\sim P(x, y)$

20. if a property is true of everything in a domain, then it is true of any particular thing in the domain

21.
$$\begin{array}{ll} \forall x, \text{ if } P(x) \text{ then } Q(x) & \forall x, \text{ if } P(x) \text{ then } Q(x) \\ P(a), \text{ for a particular } a & \sim Q(a), \text{ for a particular } a \\ \therefore \quad Q(a) & \therefore \quad \sim P(a) \end{array}$$

22. place a disk labeled A inside a disk labeled B

23. draw overlapping disks labeled A and B, respectively, and place a dot inside the part that overlaps

24. draw non-overlapping disks labeled A and B

Chapter 3: Elementary Number Theory and Methods of Proof

One aim of this chapter is to introduce you to methods for evaluating whether a given mathematical statement is true or false. Throughout the chapter the emphasis is on learning to prove and disprove statements of the form "$\forall x$ in D, if $P(x)$ then $Q(x)$." To prove such a statement directly, one supposes one has a particular but arbitrarily chosen element x in D for which $P(x)$ is true and one shows that $Q(x)$ must also be true. To disprove such a statement, one shows that there is an element x in D (a counterexample) for which $P(x)$ is true and $Q(x)$ is false. To prove such a statement by contradiction, one shows that no counterexample exists, that is, one supposes that there is an x in D for which $P(x)$ is true and $Q(x)$ is false and one shows that this supposition leads to a contradiction. Direct proof, disproof by counterexample, and proof by contradiction can, therefore, all be viewed as three aspects of one whole. A person arrives at one or the other by a thoughtful examination of the given statement, knowing what it means for a statement of that form to be true or false.

Another aim of the chapter is to help you obtain fundamental knowledge about numbers that is needed in mathematics and computer science. Note that the exercise sets contain problems of varying difficulty. Do not be discouraged if some of them are difficult for you.

Proofs given as solutions should be regarded as samples. Your instructor will probably discuss with you the particular range of proof styles that will be considered acceptable in your course.

Section 3.1

3. Think of a definition as a test. A definition says that if a certain condition is satisfied then an object can be called by a certain name. So, for instance, if an integer equals twice some integer, then it can be called even. If an integer equals twice some integer plus 1, then it can be called odd. And if an integer equals a product of two integers both of which are greater than 1, then it is composite.

   a. Yes, because $4rs = 2\cdot(2rs)$ and $2rs$ is an integer since r and s are integers and products of integers are integers.

   b. Yes, because $6r + 4s^2 + 3 = 2(3r + 2s^2 + 1) + 1$ and $3r + 2s^2 + 1$ is an integer since r and s are integers and products and sums of integers are integers.

   c. Yes, because $r^2 + 2rs + s^2 = (r+s)^2$ and $r + s$ is an integer that is greater than or equal to 2 since both r and s are positive integers and thus each is greater than or equal to 1.

6. For example, let $a = 1$ and $b = 0$. Then $\sqrt{a+b} = \sqrt{1} = 1$ and $\sqrt{a} + \sqrt{b} = \sqrt{1} + \sqrt{0} = 1$ also. Hence $\sqrt{a+b} = \sqrt{a} + \sqrt{b}$ for these values of a and b. (Note that, in fact, if a is any nonzero integer and $b = 0$, then $\sqrt{a+b} = \sqrt{a+0} = \sqrt{a} = \sqrt{a} + 0 = \sqrt{a} + \sqrt{0} = \sqrt{a} + \sqrt{b}$.)

12. *Counterexample*: Let $n = 5$. Then $\frac{n-1}{2} = \frac{5-1}{2} = \frac{4}{2} = 2$, which is not odd.

15. Note that $3n^2 - 4n + 1 = (3n - 1)(n - 1)$. Therefore, we can show that this property is true for some integers and false for other integers. For example, when $n = 2$, then $3n^2 - 4n + 1 = (3\cdot2-1)(2-1) = 5$, which is prime. However, when $n = 3$, then $3n^2 - 4n + 1 = (3\cdot3-1)(3-1) = 8 \cdot 2 = 16$, which is not prime.

18. $1^2 - 1 + 11 = 11$, which is prime. $2^2 - 2 + 11 = 13$, which is prime.
   $3^2 - 3 + 11 = 17$, which is prime. $4^2 - 4 + 11 = 23$, which is prime.
   $5^2 - 5 + 11 = 31$, which is prime. $6^2 - 6 + 11 = 41$, which is prime.
   $7^2 - 7 + 11 = 53$, which is prime. $8^2 - 8 + 11 = 67$, which is prime.
   $9^2 - 9 + 11 = 83$, which is prime. $10^2 - 10 + 11 = 101$, which is prime.

21. *Start of Proof*: Suppose x is any *[particular but arbitrarily chosen]* real number such that $x > 1$. *[We must show that $x^2 > x$.]*

27. *Proof 1*: Suppose m and n are any *[particular but arbitrarily chosen]* odd integers. *[We must show that $m+n$ is even.]* By definition of odd, there exist integers r and s such that $m = 2r+1$ and $n = 2s + 1$. Then $m + n = (2r + 1) + (2s + 1) = 2r + 2s + 2 = 2(r + s + 1)$. Let $u = r + s + 1$. Then u is an integer because r, s, and 1 are integers and a sum of integers is an integer. Hence $m + n = 2u$, where u is an integer, and so by definition of even, $m + n$ is even *[as was to be shown]*.

    Proof 2: Suppose m and n are any *[particular but arbitrarily chosen]* odd integers. *[We must show that $m+n$ is even.]* By definition of odd, there exist integers r and s such that $m = 2r+1$ and $n = 2s + 1$. Then $m + n = (2r + 1) + (2s + 1) = 2r + 2s + 2 = 2(r + s + 1)$. But $r + s + 1$ is an integer because r, s, and 1 are integers and a sum of integers is an integer. Hence $m + n$ equals twice an integer, and so by definition of even, $m + n$ is even *[as was to be shown]*.

30. *Proof*: Suppose n is any odd integer. *[We must show that $(-1)^n = -1$.]* By definition of odd, $n = 2k+1$ for some integer k. By substitution and the laws of exponents, $(-1)^n = (-1)^{2k+1} = (-1)^{2k} \cdot (-1) = ((-1)^2)^k \cdot (-1)$. But $(-1)^2 = 1$, and since 1 raised to any power equals 1, $((-1)^2)^k = 1^k = 1$. Hence, by substitution, $(-1)^n = ((-1)^2)^k \cdot (-1) = 1 \cdot (-1) = -1$ *[as was to be shown]*.

33. To prove the given statement is false, we prove that its negation is true. The negation of the statement is "For all integers k with $k \geq 4$, $2k^2 - 5k + 2$ is not prime."

    Proof of the negation: Suppose k is any integer with $k \geq 4$. *[We must show that $2k^2 - 5k + 2$ is not prime.]* We can factor $2k^2 - 5k + 2$ to obtain $2k^2 - 5k + 2 = (2k - 1)(k - 2)$. But since $k \geq 4$, $k - 2 \geq 2$. Also $2k \geq 2 \cdot 4 = 8$, and thus $2k - 1 \geq 8 - 1 = 7$. This shows that each factor of $2k^2 - 5k + 2$ is a positive integer not equal to 1, and so $2k^2 - 5k + 2$ is not prime.

42. *Proof*: Suppose m is any even integer and n is any integer. *[We must show that mn is even.]* By definition of even, there exists an integer k such that $m = 2k$. By substitution and algebra, $mn = (2k)n = 2(kn)$. But $2(kn)$ is even because kn is an integer (being a product of integers). Hence mn is even *[as was to be shown]*.

45. *Proof*: Let m and n be any odd integers. By definition of odd, $m = 2r + 1$ and $n = 2s + 1$ for some integers r and s. By substitution, $m - n = (2r + 1) - (2s + 1) = 2(r - s)$. Since $r - s$ is an integer (being a difference of integers), then $m - n$ equals twice some integer, and so $m - n$ is even by definition of even.

48. *Counterexample*: Let $m = 3$. Then $m^2 - 4 = 9 - 4 = 5$, which is not composite.

51. *Counterexample*: The number 28 cannot be expressed as a sum of three or fewer perfect squares. The only perfect squares that could be used to add up to 28 are those that are smaller than 28: 1, 4, 9, 16, and 25. The method of exhaustion can be used to show that no combination of these numbers add up to 28. (In fact, there are just three ways to express 28 as a sum of four or fewer of these numbers: $28 = 25 + 1 + 1 + 1 = 16 + 4 + 4 + 4 = 9 + 9 + 9 + 1$, and in none of these ways are only three perfect squares used.)

54. *Proof*: Suppose two consecutive integers are given. Call the smaller one n. Then the larger is $n + 1$. Let m be the difference of the squares of the numbers. Then $m = (n + 1)^2 - n^2 = (n^2 + 2n + 1) - n^2 = 2n + 1$. Because n is an integer, $m = 2 \cdot (\text{an integer}) + 1$, and so m is odd by definition of odd.

57. If m and n are perfect squares, then $m = a^2$ and $n = b^2$ for some integers a and b. We may take a and b to be nonnegative because for any real number x, $x^2 = (-x)^2$ and if x is negative then $-x$ is nonnegative. By substitution,

$$m + n + 2\sqrt{mn} = a^2 + b^2 + 2\sqrt{a^2b^2}$$
$$= a^2 + b^2 + 2ab \qquad \text{since } a \text{ and } b \text{ are nonnegative}$$
$$= (a + b)^2.$$

But $a + b$ is an integer (since a and b are), and so $m + n + 2\sqrt{mn}$ is a perfect square.

60. *a*. Note that $(x - r)(x - s) = x^2 - (r + s)x + rs$. If both r and s are odd integers, then $r + s$ is even and rs is odd (by exercises 27 and 39). If both r and s are even integers, then both $r + s$ and rs are even (by Theorem 3.1.1 and exercise 42). If one of r and s is even and the other is odd, then $r + s$ is even and rs is odd (by exercise 19 and the solution to exercise 42).

b. It follows from part(a) that $x^2 - 1253x + 255$ cannot be written as a product of the form $(x - r)(x - s)$ because for none of the possible cases (both r and s odd, both r and s even, and one of r and s odd and the other even) are both $r + s$ and rs odd integers. *[In Section 3.4, we establish formally that any integer is either even or odd.]*

Section 3.2

Additional discussion of some exercises with answers in the back of the book

- *Exercise 1*: According to the definition, if a real number can be written as a quotient of integers with a nonzero denominator, then it is rational, and if it is rational then it can be written as a quotient of integers with a nonzero denominator. In exercise 1, the number $-\frac{35}{6}$ involves a quotient of integers but it is not expressed as a quotient of integers because of the minus sign in front. To show that this number is rational, you need to find integers u and v so that $-\frac{35}{6} = \frac{u}{v}$, thereby eliminating the minus sign from the front of the number. The secret is to use the real number property: $-\frac{a}{b} = \frac{-a}{b} = \frac{a}{-b}$ (Appendix A, T12) to say that $-\frac{35}{6} = \frac{-35}{6}$.

- *Exercise 3*: Again, the problem is to express the given quantity, $\frac{4}{5} + \frac{2}{9}$, in the form $\frac{u}{v}$ where u and v are integers, and again the secret is to use one of the properties of real numbers. In this case it is that to add fractions you express each with a common denominator. (See Appendix A, T13. and T14) Thus $\frac{4}{5} + \frac{2}{9} = \frac{4 \cdot 9}{5 \cdot 9} + \frac{5 \cdot 2}{5 \cdot 9} = \frac{4 \cdot 9 + 5 \cdot 2}{5 \cdot 9} = \frac{36 + 10}{45} = \frac{46}{45}$.

Exercises Solutions

15. This statement is true. *Proof*: Suppose r and s are *[particular but arbitrarily chosen]* rational numbers. *[We must show that $r - s$ is rational.]* By definition of rational, $r = a/b$ and $s = c/d$ for some integers a, b, c, and d with $b \neq 0$ and $d \neq 0$. Then by substitution and the laws of algebra, $r - s = a/b - c/d = (ad - bc)/bd$. But $ad - bc$ and bd are both integers because a, b, c, and d are integers and products and differences of integers are integers and $bd \neq 0$ by the zero product property. Hence $r - s$ is a quotient of integers with a nonzero denominator, and so, by definition of rational number, $r - s$ is rational *[as was to be shown]*.

18. This statement is true. *Proof*: Suppose a and b are any real numbers with $a < b$. By properties T18 and T19 in Appendix A, we may add b to both sides to obtain $(a + b) < 2b$, and we may divide both sides by 2 to obtain $(a + b)/2 < b$. Similarly, since $a < b$, we may add a to both sides, which gives $2a < (a + b)$, and we may divide both sides by 2, which gives $a < (a + b)/2$. By combining the inequalities, we have $a < (a + b)/2 < b$.

21. True. *Proof*: Suppose a is any odd integer. Then $a^2 = a \cdot a$ is a product of odd integers and hence is odd by property 3. Therefore, $a^2 + a$ is a sum of odd integers and thus even by property 2.

24. *Proof*: Suppose r is any rational number. Then $r^2 = r \cdot r$ is a product of rational numbers and hence is rational by exercise 12 (or by the solution to exercise 13). Also 2 and 3, which are integers, are rational by exercise 11. Thus both $3r^2$ and $2r$ are rational by the solution to

exercise 13 (because they are products of rational numbers), and by the solution to exercise 15, $3r^2 - 2r$ is rational (because it is a difference of two rational numbers). Finally, 4, which is an integer, is rational by exercise 11. So by Theorem 3.2.2, $3r^2 - 2r + 4 = (3r^2 - 2r) + 4$ is rational. (because it is a sum of two rational numbers).

27. Yes. Since $\dfrac{ax + b}{cx + d} = 1$, then $ax + b = cx + d$, and so $ax - cx = d - b$, or, equivalently, $(a - c)x = d - b$. Thus $x = (d - b)/(a - c)$. Now $d - b$ and $a - c$ are integers because a, b, c, and d are integers and differences of integers are integers. Also $a - c \neq 0$ because it is given that $a \neq c$. Thus x can be written as a quotient of integers with a nonzero denominator, and so x is rational.

36. This incorrect proof begs the question. The second sentence asserts that a certain conclusion follows if $r + s$ is rational, and the rest of the proof uses that conclusion to deduce that $r + s$ is rational. Thus this incorrect proof assumes what is to be proved.

Section 3.3

A Note on Notation: Discrete mathematics is made up of many different mathematical topics, each with its own distinctive set of symbols. You need to know the meanings of all the symbols in the sections you cover and how to use them correctly. If you see a sentence that contains symbols and you read it out loud to yourself, does it make sense? If not, you are probably not interpreting the symbols correctly.

- Always read the equal sign as "equals" or "is equal to," and only use it between quantities that are equal. Never use the equal sign as a substitute for the word "is."

- The symbol | stands for the word "divides"; the symbol / stands for the division operator. For example, "2 | 6" is a shorthand notation for the sentence "2 divides 6," whereas 2/6 denotes the number that results from dividing 2 by 6.

- Use the $\Leftrightarrow$ symbol to substitute for the words "if, and only if" or "means that". For instance, the notation "3 | (10 + 2) $\Leftrightarrow$ 3 | 12" is a convenient shorthand for the sentence "3 divides the sum of 10 and 2 if, and only if, 3 divides 12." It would be incorrect, however, to symbolize this sentence with = in place of $\Leftrightarrow$. To see why, try reading the sentence substituting the word "equals" in place of the words "if, and only if."

Exercise Solutions

3. *Discussion*: The definition provides a test. If it is possible to express 0 as some integer times 5, then 5 divides 0. But this is possible because 0 equals 0 times 5.

   Answer: Yes, because $0 = 0 \cdot 5$.

9. Yes, because $2a \cdot 34b = 4(17ab)$ and $17ab$ is an integer since a and b are integers and $2 \mid 4$ but $a \nmid b$ because $2 \nmid 3$ and $a \nmid c$ because $2 \nmid 1$.

   and products of integers are integers.

18. *Proof*: Let m and n be any two even integers. By definition of even, $m = 2r$ and $n = 2s$ for some integers r and s. Then $mn = (2r)(2s) = 4(rs)$. Since rs is an integer (being a product of integers), mn is a multiple of 4 (by definition of divisibility).

24. *Counterexample*: Let $a = 2$, $b = 3$, and $c = 1$. Then $a \mid (b + c)$ because $2 \mid 4$ but $a \nmid b$ because $2 \nmid 3$ and $a \nmid c$ because $2 \nmid 1$.

27. *Counterexample*: Let $a = 4$ and $n = 6$. Then $a \mid n^2$ and $a \le n$ because $4 \mid 36$ and $4 \le 6$ but $a \nmid b$ because $4 \nmid 6$.

30. No. The values of nickels, dimes, and quarters are all multiples of 5. By exercise 15, a sum of numbers divisible by 5 is also divisible by 5. So since \$4.72 is not a multiple of 5, \$4.72 cannot be obtained using only nickels, dimes, and quarters.

33. *b*. Let $N = 12,858,306,120,312$. The sum of the digits of N is 42, which is divisible by 3 but not by 9. Therefore, N is divisible by 3 but not by 9. The right-most digit of N is neither 5 nor 0, and so N is not divisible by 5. The two right-most digits of N are 12, which is divisible by 4. Therefore, N is divisible by 4.

c. Let $N = 517,924,440,926,512$. The sum of the digits of N is 61, which is not divisible by 3 (and hence not by 9 either). Therefore, N is not divisible either by 3 or by 9. The right-most digit of N is neither 5 nor 0, and so N is not divisible by 5. The two right-most digits of N are 12, which is divisible by 4. Therefore, N is divisible by 4.

d. Let $N = 14,328,083,360,232$. The sum of the digits of N is 45, which is divisible by 9 and hence also by 3. Therefore, N is divisible by 9 and by 3. The right-most digit of N is neither 5 nor 0, and so N is not divisible by 5. The two right-most digits of N are 32, which is divisible by 4. Therefore, N is divisible by 4.

36. *a.* $p_1^{3e_1} \cdot p_2^{3e_2} \cdots p_k^{3e_k}$

b. $k = 2^2 \cdot 3 \cdot 7^2 \cdot 11$

product $= 2^4 \cdot 3^5 \cdot 7 \cdot 11^2 \cdot k = 2^6 \cdot 3^6 \cdot 7^3 \cdot 11^3 = (2^2 \cdot 3^2 \cdot 7 \cdot 11)^3 = 2772^3$

39. *b.*

$$
\begin{aligned}
20! &= 20 \cdot 19 \cdot 18 \cdot 17 \cdot 16 \cdot 15 \cdot 14 \cdot 13 \cdot 12 \cdot 11 \cdot 10 \cdot 9 \cdot 8 \cdot 7 \cdot 6 \cdot 5 \cdot 4 \cdot 3 \cdot 2 \cdot 1 \\
&= 2^2 \cdot 5 \cdot 19 \cdot 2 \cdot 3^2 \cdot 17 \cdot 2^4 \cdot 3 \cdot 5 \cdot 2 \cdot 7 \cdot 13 \cdot 2^2 \cdot 3 \cdot 11 \cdot 2 \cdot 5 \cdot 3^2 \cdot 2^3 \cdot 7 \cdot 2 \cdot 3 \cdot 5 \cdot 2^2 \cdot 3 \cdot 2 \\
&= 2^{18} \cdot 3^8 \cdot 5^4 \cdot 7^2 \cdot 11 \cdot 13 \cdot 17 \cdot 19
\end{aligned}
$$

c. Squaring the result of part (b) gives

$$
\begin{aligned}
(20!)^2 &= (2^{18} \cdot 3^8 \cdot 5^4 \cdot 7^2 \cdot 11 \cdot 13 \cdot 17 \cdot 19)^2 \\
&= 2^{36} \cdot 3^{16} \cdot 5^8 \cdot 7^4 \cdot 11^2 \cdot 13^2 \cdot 17^2 \cdot 19^2
\end{aligned}
$$

When $(20!)^2$ is written in ordinary decimal form, there are as many zeros at the end of it as there are factors of the form $2 \cdot 5 \, (= 10)$ in its prime factorization. Thus, since the prime factorization of $(20!)^2$ contains eight 5's and more than eight 2's, $(20!)^2$ contains eight factors of 10 and hence eight zeros.

42. *Proof*: Suppose n is a nonnegative integer whose decimal representation ends in 5. By the hint for exercise 41, $n = 10m + 5$ for some integer m. By factoring out a 5, $n = 10m + 5 = 5(2m + 1)$, and $2m + 1$ is an integer since m is an integer. Hence n is divisible by 5.

45. *Proof*: Suppose n is any nonnegative integer for which the sum of the digits of n is divisible by 3. By the same reasoning as in the answer to exercise 44,

$$
\begin{aligned}
n &= 9(d_k \cdot \underbrace{11\ldots1}_{k\ 1\text{'s}} + d_{k-1} \cdot \underbrace{11\ldots1}_{(k-1)\ 1\text{'s}} + \cdots + d_2 \cdot 11 + d_1) + (d_k + d_{k-1} + \cdots + d_2 + d_1 + d_0) \\
&= 3[3(d_k \cdot \underbrace{11\ldots1}_{k\ 1\text{'s}} + d_{k-1} \cdot \underbrace{11\ldots1}_{(k-1)\ 1\text{'s}} + \cdots + d_2 \cdot 11 + d_1)] + (d_k + d_{k-1} + \cdots + d_2 + d_1 + d_0) \\
&= (\text{an integer divisible by 3}) + (\text{the sum of the digits of } n).
\end{aligned}
$$

Since the sum of the digits of n is divisible by 3, n can be written as a sum of two integers each of which is divisible by 3. It follows from exercise 15 that n is divisible by 3.

Section 3.4

6. $q = -4, r = 5$

9. *a.* 5 *b.* 3

12. Let the days of the week be numbered from 0 (Sunday) through 6 (Saturday) and let $DayT$ and $DayN$ be variables representing the day of the week today and the day of the week N days from today. By the quotient-remainder theorem, there exist unique integers q and r such that $DayT + N = 7q + r$ and $0 \le r < 7$. Now $DayT + N$ counts the number of days to the day N days from today starting last Sunday (where "last Sunday" is interpreted to mean today if today is a Sunday). Thus $DayN$ is the day of the week that is $DayT + N$ days from last Sunday. Because each week has seven days, $DayN$ is the same as the day of the week $DayT + N - 7q$ days from last Sunday. But $DayT + N - 7q = r$ and $0 \le r < 7$. Therefore, $DayN = r = (DayT + N) \bmod 7$.

15. There are 13 leap year days between January 1, 2000 and January 1, 2050 (once every four years in 2000, 2004, 2008, 2012, . . . , 2048). So 13 of the years have 366 days and the remaining 38 years have 365 days. This gives a total of $13 \cdot 366 + 37 \cdot 365 = 18,263$ days between the two dates. Using the formula $DayN = (DayT + N) \bmod 7$, and letting $DayT = 6$ (Saturday) and $N = 18,263$ gives $DayN = (6 + 18263) \bmod 7 = 18269 \bmod 7 = 6$, which is also a Saturday.

18. When b is divided by 12, the remainder is 5. Thus there exists an integer m so that $b = 12m + 5$. Multiplying this equation by 8 gives $8b = 96m + 40 = 96m + 36 + 4 = 12(8m + 4) + 4$. Since $8m + 4$ is an integer and since $0 \le 4 < 12$, the uniqueness part of the quotient-remainder theorem guarantees that the remainder obtained when $8b$ is divided by 12 is 4.

21. Recall that (1) *A is a sufficient condition for B* means *if A then B*, and (2) *A is a necessary condition for B* means *if B then A*. Thus proving the given statement requires proving both a (universal) conditional statement and its converse.

    Proof: ($\Rightarrow$) *[We first prove that given any nonnegative integer n and positive integer d, if n is divisible by d, then n mod d = 0.]* Suppose n is any nonnegative integer and d is any positive integer such that n is divisible by d. By definition of divisibility, $n = dk$ for some integer k. Thus the equation $n = dk + 0$ is true and the inequality $0 \le 0 < d$ is also true, and so, by the uniqueness part of the quotient-remainder theorem, $n \bmod d = 0$.

    ($\Leftarrow$) *[Second, we prove the converse, namely that given any nonnegative integer n and positive integer d, if n mod d = 0 then n is divisible by d.]* Suppose n is any nonnegative integer and d is any positive integer such that $n \bmod d = 0$. Then 0 is the remainder obtained when n is divided by d, and thus by the quotient-remainder theorem there exists an integer q such that $n = dq + 0$. But this is equivalent to $n = dq$, and so, by definition of divisibility, n is divisible by d.

24. *Proof:* Consider any two consecutive integers. Call the smaller one n. By the quotient-remainder theorem with $d = 2$, either n is even or n is odd.

    Case 1 (n is even): In this case $n = 2k$ for some integer k. Then $n(n + 1) = 2k(2k + 1) = 2[k(2k + 1)]$. But $k(2k + 1)$ is an integer (because products and sums of integers are integers), and so $n(n + 1)$ is even.

    Case 2 (n is odd): In this case $n = 2k + 1$ for some integer k. Then

    $$n(n + 1) = (2k + 1)[(2k + 1) + 1] = (2k + 1)(2k + 2) = 2[(2k + 1)(k + 1)].$$

    But $(2k + 1)(k + 1)$ is an integer (because products and sums of integers are integers), and so $n(n + 1)$ is even.

    Hence in either case the product $n(n + 1)$ is even *[as was to be shown]*.

30. *Proof:* Suppose n and $n + 1$ are any two consecutive integers. By the quotient-remainder theorem with $d = 3$, we know that $n = 3q$, or $n = 3q + 1$, or $n = 3q + 2$ for some integer q.

Case 1 ($n = 3q$ for some integer q): In this case, $n(n + 1) = 3q(3q + 1) = 3[q(3q + 1)]$. Let $k = q(3q + 1)$. Then k is an integer because sums and products of integers are integers. Hence $n(n + 1) = 3k$ for some integer k.

Case 2 ($n = 3q + 1$ for some integer q): In this case, $n(n+1) = (3q+1)(3q+2) = 9q^2+9q+2 = 3(3q^2 + 3q) + 2$. Let $k = 3q^2 + 3q$. Then k is an integer because sums and products of integers are integers. Hence $n(n + 1) = 3k + 2$ for some integer k.

Case 3 ($n = 3q + 2$ for some integer q): In this case, $n(n + 1) = n(3q + 3) = 3[n(q + 1)]$. Let $k = n(q + 1)$. Then k is an integer because sums and products of integers are integers. Hence $n(n + 1) = 3k$ for some integer k.

Thus in all three cases, the product of the two consecutive integers either equals $3k$ or it equals $3k + 2$ for some integer k *[as was to be shown]*.

33. *Proof:* Suppose a, b, and c are any integers such that $a - b$ is odd and $b - c$ is even. Then $(a - b) + (b - c)$ is a sum of an odd integer and an even integer and hence is odd (by Example 3.2.3 #5). But $(a - b) + (b - c) = a - c$, and thus $a - c$ is odd.

36. *Proof:* Suppose n is any integer. *[We must show that $8 \mid n(n+1)(n+2)(n+3)$.]* By the quotient-remainder theorem with $d = 4$, $n = 4k$ or $n = 4k + 1$ or $n = 4k + 2$ or $n = 4k + 3$ for some integer k.

Case 1 ($n = 4k$ for some integer k): In this case,

$$n(n + 1)(n + 2)(n + 3) = 4k(4k + 1)(4k + 2)(4k + 3) = 8[k(4k + 1)(2k + 1)(4k + 3)],$$

which is divisible by 8 (because k is an integer and sums and products of integers are integers).

Case 2 ($n = 4k + 1$ for some integer k): In this case,

$$n(n + 1)(n + 2)(n + 3) = (4k + 1)(4k + 2)(4k + 3)(4k + 4) = 8[(4k + 1)(2k + 1)(4k + 3)(k + 1)],$$

which is divisible by 8 (because k is an integer and sums and products of integers are integers).

Case 3 ($n = 4k + 2$ for some integer k): In this case,

$$n(n + 1)(n + 2)(n + 3) = (4k + 2)(4k + 3)(4k + 4)(4k + 5) = 8[(2k + 1)(4k + 3)(k + 1)(4k + 5)],$$

which is divisible by 8 (because k is an integer and sums and products of integers are integers).

Case 4 ($n = 4k + 3$ for some integer k): In this case,

$$n(n + 1)(n + 2)(n + 3) = (4k + 3)(4k + 4)(4k + 5)(4k + 6) = 8[(4k + 3)(k + 1)(4k + 5)(2k + 3)],$$

which is divisible by 8 (because k is an integer and sums and products of integers are integers).

Hence in all four possible cases, $8 \mid n(n + 1)(n + 2)(n + 3)$ *[as was to be shown]*.

39. *Proof:* Consider any four consecutive integers. Call the smallest n. Then the sum of the four integers is $n + (n + 1) + (n + 2) + (n + 3) = 4n + 6 = 4(n + 1) + 2$. Let $k = n + 1$. Then k is an integer because it is a sum of integers. Hence n can be written in the required form.

42. *Proof:* Let p be any prime number except 2 or 3. By the quotient-remainder theorem, p can be written as $6k$ or $6k + 1$ or $6k + 2$ or $6k + 3$ or $6k + 4$ or $6k + 5$ for some integer k. Since p is prime and $p \neq 2$, p is not divisible by 2. Consequently, $p \neq 6k$, $p \neq 6k + 2$, and $p \neq 6k + 4$ for any integer k *[because all of these numbers are divisible by 2]*. Furthermore, since p is prime and $p \neq 3$, p is not divisible by 3. Thus $p \neq 6k + 3$ *[because this number is divisible by 3]*. Therefore, $p = 6k + 1$ or $p = 6k + 5$ for some integer k.

is another significant part of the conclusion suggests operating on the inequality $n < x < n+1$ to make the middle term $m - x$. Doing so gives information about the floor of $m - x$, which then makes it possible to use the definition of floor to draw the required conclusion.

Proof: Suppose m is any integer and x is any real number that is not an integer. By definition of floor, $\lfloor x \rfloor = n$ where n is an integer and $n \le x < n+1$. Since x is not an integer, $x \ne n$, and so $n < x < n+1$. Multiply all parts of this inequality by -1 to obtain $-n > -x > -n - 1$. Then add m to all parts to obtain $m - n > m - x > m - n - 1$, or, equivalently, $m - n - 1 < m - x < m - n$. But $m - n - 1$ and $m - n$ are both integers, and so by definition of floor, $\lfloor m - x \rfloor = m - n - 1$. By substitution, $\lfloor x \rfloor + \lfloor m - x \rfloor = n + (m - n - 1) = m - 1$ *[as was to be shown].*

27. *Proof:* Suppose x is any real number such that $x - \lfloor x \rfloor \ge 1/2$. Multiply both sides by 2 to obtain $2x - 2\lfloor x \rfloor \ge 1$, or equivalently, $2x \ge 2\lfloor x \rfloor + 1$. Now by definition of floor, $x < \lfloor x \rfloor + 1$. Hence $2x < 2\lfloor x \rfloor + 2$. Put the two inequalities involving x together to obtain $2\lfloor x \rfloor + 1 \le 2x < 2\lfloor x \rfloor + 2$. By definition of floor, then, $\lfloor 2x \rfloor = 2\lfloor x \rfloor + 1$.

Section 3.6

6. *Negation of statement:* There is a greatest negative real number.

Proof of statement: Suppose not. That is, suppose there is a greatest negative real number a. *[We must deduce a contradiction.]* Then $a < 0$ and $a \ge x$ for every negative real number x. Let $b = a/2$. Then b is a real number because b is a quotient of two real numbers (with a nonzero denominator). Also $a < a/2 < 0$. *[The reason is that $0 < 1/2 < 1$ and multiplying all parts by a, which is less than zero, gives $a < a/2 < 0$.]* By substitution, then, $a < b < 0$. Thus b is a negative real number that is greater than a. This contradicts the supposition that a is the greatest negative real number. *[Hence the supposition is false and the statement is true.]*

9. *Proof:* Suppose not. That is, suppose there are real numbers x and y such that x is irrational, y is rational and $x - y$ is rational. *[We must derive a contradiction.]* By definition of rational, $y = a/b$ and $x - y = c/d$ for some integers a, b, c, and d with $b \ne 0$ and $d \ne 0$. Then, by substitution, $x - \frac{a}{b} = \frac{c}{d}$. Solve this equation for x to obtain $x = \frac{c}{d} + \frac{a}{b} = \frac{bc}{bd} + \frac{ad}{bd} = \frac{bc+ad}{bd}$. But both $bc + ad$ and bd are integers because products and sums of integers are integers, and $bd \ne 0$ by the zero product property. Hence x is a ratio of integers with a nonzero denominator, and so x is rational by definition of rational. This contradicts the supposition that x is irrational. *[Hence the supposition is false, and the given statement is true.]*

12. *Proof 1:* Suppose not. That is, suppose $\exists$ an integer n such that $4 \mid (n^2 - 2)$. *[We must derive a contradiction.]* By definition of divisibility, $n^2 - 2 = 4m$ for some integer m. By the quotient-remainder theorem with $d = 2$, n is either even or odd.

Case 1 (n is even): In this case, $n = 2k$ for some integer k, and so, by substitution, $n^2 - 2 = (2k)^2 - 2 = 4k^2 - 2 = 4m$, where m is an integer. Thus $4k^2 - 4m = 2$, and hence $k^2 - m = \frac{1}{2}$. But the left-hand side of this equation is an integer (because k and m are integers) and the right-hand side is not an integer. Since this is impossible, the case where n is even cannot occur.

Case 2 (n is odd): In this case, $n = 2k + 1$ for some integer k, and so, by substitution, $n^2 - 2 = (2k+1)^2 - 2 = 4k^2 + 4k + 1 - 2 = 4k^2 + 4k - 1 = 4m$, where m is an integer. Thus $4k^2 + 4k - 4m = 1$, and hence $k^2 + k - m = \frac{1}{4}$. But the left-hand side of this equation is an integer (because k and m are integers) and the right-hand side is not an integer. Since this is impossible, the case where n is odd cannot occur.

It follows from cases 1 and 2 that n can be neither even nor odd, which contradicts the fact that it must be one or the other. *[Therefore the supposition is false, and the given statement is true.]*

Proof 2: Suppose not. That is, suppose $\exists$ an integer n such that $4 \mid (n^2 - 2)$. *[We must derive a contradiction.]* By definition of divisibility, $n^2 - 2 = 4m$ for some integer m. Then $n^2 = 4m + 2 = 2(2m + 1)$, and so n^2 is even. Hence n is even (by Proposition 3.6.4). By definition of even, $n = 2k$ for some integer k. Substituting into the equation $n^2 - 2 = 4m$ gives $(2k)^2 - 2 = 4k^2 - 2 = 4m$, and so $4k^2 = 4m + 2$. Dividing by 2 gives $2k^2 = 2m + 1$. Since k^2 is an integer, this implies that $2m + 1$ is even, but, since m is an integer, $2m + 1$ is odd. This contradicts Theorem 3.6.2 that no integer is both even and odd . *[Hence the supposition is false and the statement is true.]*

15. *Idea of Proof*: This proof goes by contradiction, and so it begins by assuming that there are some odd integers and a rational number with a certain property. To arrive at a contradiction, the proof uses a similar strategy as is used for the proof of the irrationality of $\sqrt{2}$, namely the assumption that the fraction is first reduced to lowest terms so that the the numerator and denominator have no common factors greater than 1. The most important algebraic step is to multiply both sides of the equation by the greatest denominator in the resulting equation to obtain an expression in integers that is equal to zero. From that point, four possible cases are considered, depending on whether the numerator and denominator of the fraction are even or odd, and in each case, a contradiction is obtained.

Proof: Suppose not. That is, suppose $\exists$ odd integers a, b, and c and a rational number z such that $az^2 + bz + c = 0$. By definition of rational, there exist integers m and n such that $z = m/n$ and $n \neq 0$. By cancelling common factors if necessary, we may assume that m and n have no common factors. By substitution, $a(\frac{m}{n})^2 + b(\frac{m}{n}) + c = 0$, and multiplying both sides by n^2 gives $am^2 + bmn + cn^2 = 0$.

Case 1 (both m and n are even): In this case both m and n have a factor of 2, which contradicts the assumption that m and n have no common factors.

Case 2 (m is odd and n is even): In this case am^2 is a product of odd integers, and hence is odd (by Example 3.2.3 #3), and both bmn and cn^2 are products that contain an even factor, and hence are even (by Example 3.2.3 #1 and #4). Thus $bmn + cn^2$ is a sum of even integers, which is even (by Example 3.2.3 #1), and so $am^2 + bmn + cn^2$ is the sum of an odd integer and an even integer, which is odd (by Example 3.2.3 #5). But $am^2 + bmn + cn^2 = 0$, which is even. Therefore, $am^2 + bmn + cn^2$ is both even and odd, which contradicts Theorem 3.6.2.

Case 3 (m is even and n is odd): In this case cn^2 is a product of odd integers, and hence is odd (by Example 3.2.3 #3), and both am^2 and bmn are products that contain an even factor, and hence are even (by Example 3.2.3 #1 and #4). Thus $am^2 + bmn$ is a sum of even integers, which is even (by Example 3.2.3 #1), and so $am^2 + bmn + cn^2$ is the sum of an even integer and an odd integer, which is odd (by Example 3.2.3 #5). But $am^2 + bmn + cn^2 = 0$, which is even. Therefore, $am^2 + bmn + cn^2$ is both even and odd, which contradicts Theorem 3.6.2.

Case 4 (both m and n are odd): In this case all three products am^2, bmn, and cn^2 consist only of odd factors, and hence all are odd (by Example 3.2.3 #3). Thus $am^2 + bmn$ is the sum of two odd integers, which is even (by Example 3.2.3 #2), and so $am^2 + bmn + cn^2$ is the sum of an even and an odd integer, which is odd (by Example 3.2.3 #5). But $am^2 + bmn + cn^2 = 0$, which is even. Therefore, $am^2 + bmn + cn^2$ is both even and odd, which contradicts Theorem 3.6.2.

Hence in all four cases a contradiction is reached, and so the supposition is false and the given statement is true.

18. *Proof (by contraposition)*: Suppose a and b are *[particular but arbitrarily chosen]* real numbers such that $a \geq 25$ and $b \geq 25$. Then $a + b \geq 25 + 25 = 50$. Hence if $a + b < 50$, then $a < 25$ or $b < 25$.

30. After crossing out all multiples of 2, 3, 5, and 7 (the prime numbers less than $\sqrt{100}$), the remaining numbers are prime. They are circled in the following diagram.

(2) (3) 4̸ (5) 6̸ (7) 8̸ 9̸ 1̸0̸ (11) 1̸2̸ (13) 1̸4̸ 1̸5̸

1̸6̸ (17) 1̸8̸ (19) 2̸0̸ 2̸1̸ 2̸2̸ (23) 2̸4̸ 2̸5̸ 2̸6̸ 2̸7̸ 2̸8̸ (29)

3̸0̸ (31) 3̸2̸ 3̸3̸ 3̸4̸ 3̸5̸ 3̸6̸ (37) 3̸8̸ 3̸9̸ 4̸0̸ (41) 4̸2̸ (43)

4̸4̸ 4̸5̸ 4̸6̸ (47) 4̸8̸ 4̸9̸ 5̸0̸ 5̸1̸ 5̸2̸ (53) 5̸4̸ 5̸5̸ 5̸6̸ 5̸7̸

5̸8̸ (59) 6̸0̸ (61) 6̸2̸ 6̸3̸ 6̸4̸ 6̸5̸ 6̸6̸ (67) 6̸8̸ 6̸9̸ 7̸0̸ (71)

7̸2̸ (73) 7̸4̸ 7̸5̸ 7̸6̸ 7̸7̸ 7̸8̸ (79) 8̸0̸ 8̸1̸ 8̸2̸ (83) 8̸4̸ 8̸5̸

8̸6̸ 8̸7̸ 8̸8̸ (89) 9̸0̸ 9̸1̸ 9̸2̸ 9̸3̸ 9̸4̸ 9̸5̸ 9̸6̸ (97) 9̸8̸ 9̸9̸

Section 3.7

6. False.

Proof 1: $\sqrt{2}/6 = (1/6) \cdot \sqrt{2}$, which is a product of a nonzero rational number and an irrational number. By exercise 10 of Section 3.6, such a product is irrational.

Proof 2: Suppose not. Suppose $\sqrt{2}/6$ is rational. *[We must derive a contradiction.]* By definition of rational, $\exists$ integers a and b with $\sqrt{2}/6 = a/b$ and $b \neq 0$. Solving for $\sqrt{2}$ gives $\sqrt{2} = 6a/b$. But $6a$ is an integer (because products of integers are integers) and b is a nonzero integer. Therefore by definition of rational, $\sqrt{2}$ is rational. This contradicts Theorem 3.7.1 which states that $\sqrt{2}$ is irrational.

12. *Counterexample*: $\sqrt{2}$ is irrational. Also $\sqrt{2} \cdot \sqrt{2} = 2$ and 2 is rational because $2 = 2/1$. Thus $\exists$ irrational numbers whose product is rational.

15. a. *Proof (by contraposition)*: Let n be any integer such that n is not even. *[We must show that n^3 is not even.]* By Theorem 3.6.2 n is odd, and so n^3 is also odd (by Example 3.2.3 (#3) applied twice). Thus (again by Theorem 3.6.2), n^3 is not even *[as was to be shown]*.

b. *Proof*: Suppose not. That is, suppose $\sqrt[3]{2}$ is rational. *[We must derive a contradiction.]* By definition of rational, $\sqrt[3]{2} = a/b$ for some integers a and b with $b \neq 0$. By cancelling any common factors if necessary, we may assume that a and b have no common factors. Cubing both sides of the equation $\sqrt[3]{2} = a/b$ gives $2 = a^3/b^3$, and so $2b^3 = a^3$. Thus a^3 is even. By part (a) of this questions, a is even, and thus $a = 2k$ for some integer k. By substitution $a^3 = (2k)^3 = 8k^3 = 2b^3$, and so $b^3 = 4k^3 = 2(2k^3)$. It follows that b^3 is even, and hence (also by part (a)) b is even. Thus both a and b are even which contradicts the assumption that a and b have no common factor. Therefore, the supposition is false, and $\sqrt[3]{2}$ is irrational.

18. *Proof*: Suppose that a and d are integers with $d > 0$ and that q_1, q_2, r_1, and r_2 are integers such that $a = dq_1 + r_1$ and $a = dq_2 + r_2$, where $0 \leq r_1 < d$ and $0 \leq r_2 < d$. *[We must show that $r_1 = r_2$ and $q_1 = q_2$.]* Then $dq_1 + r_1 = dq_2 + r_2$, and so $r_2 - r_1 = dq_1 - dq_2 = d(q_1 - q_2)$. This implies that $d \mid (r_2 - r_1)$ because $q_1 - q_2$ is an integer (being a difference of integers). But both r_1 and r_2 lie between 0 and d. Thus the difference of $r_2 - r_1$ lies between $-d$ and d. *[For, by properties T22 and T25 of Appendix A, if $0 \leq r_1 < d$ and $0 \leq r_2 < d$, then multiplying the first inequality by -1 gives $0 \geq -r_1 > -d$ or, equivalently, $-d < -r_1 \leq 0$. And adding $-d < -r_1 \leq 0$ and $0 \leq r_2 < d$ gives $-d < r_2 - r_1 < d$.]* Since $r_2 - r_1$ is a multiple of d and yet lies between $-d$ and d, the only possibility is that $r_2 - r_1 = 0$, or, equivalently, that $r_1 = r_2$. Substituting back into the original expressions for a and equating the two gives $dq_1 + r_1 = dq_2 + r_1$ [because $r_1 = r_2$]. Subtracting r_1 from both sides gives $dq_1 = dq_2$, and since $d \neq 0$, we have that $q_1 = q_2$. Hence, $r_1 = r_2$ and $q_1 = q_2$, as was to be shown.

21. *b. Proof:* Suppose not. That is, suppose $\sqrt{2}$ is rational. *[We will show that this supposition leads to a contradiction.]* By definition of rational, we may write $\sqrt{2} = a/b$ for some integers a and b with $b \neq 0$. Then $2 = a^2/b^2$, and so $a^2 = 2b^2$. Consider the prime factorizations for a^2 and for $2b^2$. By the unique factorization theorem, these factorizations are unique except for the order in which the factors are written down. Now because every prime factor of a occurs twice in the prime factorization of a^2, the prime factorization of a^2 contains an even number of 2's. If 2 is a factor of a, then this even number is positive, and if 2 is not a factor of a, then this even number is 0. On the other hand, because every prime factor of b occurs twice in the prime factorization of b^2, the prime factorization of $2b^2$ contains an odd number of 2's. Therefore, the equation $a^2 = 2b^2$ cannot be true. So the supposition is false, and hence $\sqrt{2}$ is irrational.

24. *Proof:* Suppose not. That is, suppose that $\log_5(2)$ is rational. *[We will show that this supposition leads to a contradiction.]* By definition of rational, $\log_5(2) = a/b$ for some integers a and b with $b \neq 0$. Since logarithms are always positive, we may assume that a and b are both positive. By definition of logarithm, $5^{a/b} = 2$. Raising both sides to the bth power gives $5^a = 2^b$. Let $N = 5^a = 2^b$. Since $b \geq 0$, $N > 2^0 = 1$. Thus we may consider the prime factorization of N. Because $N = 5^a$, the prime factors of N are all 5. On the other hand, because $N = 2^b$, the prime factors of N are all 2. This contradicts the unique factorization theorem which states that the prime factors of any integer greater than 1 are unique except for the order in which they are written. Hence the supposition is false, and so $\log_5(2)$ is irrational.

27. *a.* The following are all prime numbers: $N_1 = 2 + 1 = 3$, $N_2 = 2\cdot3 + 1 = 7$, $N_3 = 2\cdot3\cdot5 + 1 = 31$, $N_4 = 2\cdot3\cdot5\cdot7 + 1 = 211$, $N_5 = 2\cdot3\cdot5\cdot7\cdot11 + 1 = 2311$. However, $N_6 = 2\cdot3\cdot5\cdot7\cdot11\cdot13 + 1 = 30031 = 59\cdot509$. Thus the smallest non-prime integer of the given form is 30,031.

b. Each of N_1, N_2, N_3, N_4, and N_5 is prime, and so each is its own smallest prime divisor. Thus $q_1 = N_1$, $q_2 = N_2$, $q_3 = N_3$, $q_4 = N_4$, and $q_5 = N_5$. However, N_6 is not prime and $N_6 = 30031 = 59\cdot509$. Since 59 and 509 are primes, the smallest prime divisor of N_6 is $q_6 = 59$.

30. *Proof:* Suppose n is any integer that is greater than 2. Then $n! - 1$ is an integer that is greater than 2, and so by Theorem 3.3.2 there is a prime number p that divides $n! - 1$. Thus $p \leq n! - 1 < (n!)$. Now either $p > n$ or $p \leq n$. Suppose $p \leq n$. *[We will show that this supposition leads to a contradiction.]* Because $p \leq n$, then $p \mid (n!)$. So $p \mid (n!)$ and also $p \mid (n! - 1)$, and thus (by exercise 16 of Section 3.3) p divides $(n! - (n! - 1))$, which equals 1. But the only divisors of 1 are 1 and -1, and so $p = 1$ or $p = -1$. However since p is prime, $p > 1$. Thus we have reached a contradiction. Hence the supposition that $p \leq n$ is false, and so $p > n$. Therefore, p is a prime number such that $n < p < (n!)$.

33. *Existence Proof:* When $n = 2$, then $n^2 + 2n - 3 = 2^2 + 2 \cdot 2 - 3 = 5$, which is prime. Thus there is a prime number of the form $n^2 + 2n - 3$, where n is a positive integer.

Uniqueness Proof (by contradiction): Suppose not. By the existence proof above, we know that when $n = 2$, then $n^2 + 2n - 3$ is prime. Suppose there is another positive integer m, not equal to 2, such that $m^2 + 2m - 3$ is prime. *[We must derive a contradiction.]* By factoring, we see that $m^2 + 2m - 3 = (m + 3)(m - 1)$. Now $m \neq 1$ because otherwise $m^2 + 2m - 3 = 0$, which is not prime. Also $m \neq 2$ by supposition. Thus $m > 2$. Consequently, $m + 3 > 5$ and $m - 1 > 1$, and so $m^2 + 2m - 3$ can be written as a product of two positive integers neither of which is 1 (namely $m + 3$ and $m - 1$). This contradicts the supposition that $m^2 + 2m - 3$ is prime. Hence the supposition is false: there is no integer m other than 2 such that $m^2 + 2m - 3$ is prime.

Uniqueness Proof (direct): Suppose m is any positive integer such that $m^2 + 2m - 3$ is prime. *[We must show that $m = 2$.]* By factoring, $m^2 + 2m - 3 = (m + 3)(m - 1)$. Since $m^2 + 2m - 3$ is prime, either $m + 3 = 1$ or $m - 1 = 1$. Now $m + 3 \neq 1$ because m is positive (and if $m + 3 = 1$ then $m = -2$). Thus $m - 1 = 1$, which implies that $m = 2$ *[as was to be shown]*.

Section 3.8

3. *b.* $z = 6$

12. *Solution 1:* $\gcd(48, 54) = \gcd(6 \cdot 8, 6 \cdot 9) = 6$

 Solution 2: $\gcd(48, 54) = \gcd(2^4 \cdot 3, 2 \cdot 3^3) = 2 \cdot 3 = 6$

15.

$$
\begin{array}{r}
13 \\
832\overline{)10933} \\
10816 \\
\hline
117
\end{array}
$$
So $10933 = 832 \cdot 13 + 117$, and hence $\gcd(10933, 832) = \gcd(832, 117)$

$$
\begin{array}{r}
7 \\
117\overline{)832} \\
819 \\
\hline
13
\end{array}
$$
So $832 = 117 \cdot 7 + 13$, and hence $\gcd(832, 117) = \gcd(117, 13)$

$$
\begin{array}{r}
9 \\
13\overline{)117} \\
117 \\
\hline
0
\end{array}
$$
So $117 = 13 \cdot 9 + 0$, and hence $\gcd(117, 13) = \gcd(13, 0)$

But $\gcd(13, 0) = 13$. So $\gcd(10933, 832) = 13$.

18.

A	5859						
B	1232						
r	1232	931	301	28	21	7	0
a	5859	1232	931	301	28	21	7
b	1232	931	301	28	21	7	0
gcd							7

21. *Proof:* Suppose a and b are any integers with $b \neq 0$, and suppose q and r are any integers such that $a = bq + r$. *[We must show that $\gcd(b, r) \leq \gcd(a, b)$.]*

Step 1 (proof that any common divisor of b and r is also a common divisor of a and b): Let c be a common divisor of b and r. Then $c \mid b$ and $c \mid r$, and so by definition of divisibility, $b = nc$ and $r = mc$ for some integers n and m. Now substitute into the equation $a = bq + r$ to obtain $a = (nc)q + mc = c(nq + m)$. But $nq + m$ is an integer, and so by definition of divisibility $c \mid a$. Now we already know that $c \mid b$; hence c is a common divisor of a and b.

Step 2 (proof that $\gcd(b, r) \leq \gcd(a, b)$): By step 1, every common divisor of b and r is a common divisor of a and b. It follows that the greatest common divisor of b and r is a common divisor of a and b. But then $\gcd(b, r)$ (being one of the common divisors of a and b) is less than or equal to the greatest common divisor of a and b: $\gcd(b, r) \leq \gcd(a, b)$.

24. *a. Proof:* Suppose a and b are integers and $a \geq b > 0$. *[We first show that every common divisor of a and b is a common divisor of b and $a - b$, and conversely.]*

Part 1 (proof that every common divisor of a and b is a common divisor of b and $a - b$):

Suppose $d \mid a$ and $d \mid b$. Then $d \mid (a - b)$ by exercise 14 of Section 3.3. Hence d is a common divisor of a and $a - b$.

Part 2 (proof that every common divisor of b and a − b is a common divisor of a and b):

Suppose $d \mid b$ and $d \mid (a - b)$. Then by exercise 16 of Section 3.3, $a \mid [b + (a - b)]$. But $b + (a - b) = a$, and so $d \mid a$. Hence d is a common divisor of a and b.

Part 3 (end of proof): Because every common divisor of a and b is a common divisor of b and $a - b$, the greatest common divisor of a and b is a common divisor of b and $a - b$ and so is less than or equal to the greatest common divisor of a and $a - b$. Thus $\gcd(a, b) \le \gcd(b, a - b)$. By similar reasoning, $\gcd(b, a - b) \le \gcd(a, b)$. Therefore, $\gcd(a, b) = \gcd(b, a - b)$.

c.

A	768											
B	348											
a	768	420	72					12				0
b	348			276	204	132	60		48	36	24	12
gcd												12

27. *Proof:* Let a and b be any positive integers.

Part 1 (proof that if $\operatorname{lcm}(a, b) = b$ then $a \mid b$): Suppose that $\operatorname{lcm}(a, b) = b$. By definition of least common multiple, $a \mid \operatorname{lcm}(a, b)$, and so by substitution, $a \mid b$.

Part 2 (proof that if $a \mid b$ then $\operatorname{lcm}(a, b) = b$): Suppose that $a \mid b$. Then since it is also the case that $b \mid b$, b is a common multiple of a and b. Moreover, because b divides any common multiple of both a and b, $\operatorname{lcm}(a, b) = b$.

General Review Guide: Chapter 3

Definitions

- Why is the phrase "if, and only if" used in a definition? *(p. 127)*
- How are the following terms defined?
  - even integer *(p. 127)*
  - odd integer *(p. 127)*
  - prime number *(p. 128)*
  - composite number *(p. 128)*
  - rational number *(p. 141)*
  - divisibility of one integer by another *(p. 148)*
  - the floor of a real number *(p. 165)*
  - the ceiling of a real number *(p. 165)*
  - greatest common divisor of two integers *(p. 192)*

Proving an Existential Statement/Disproving a Universal Statement

- How do you determine the truth of an existential statement? *(p. 128)*
- What does it mean to "disprove" a statement? *(p. 129)*
- What is disproof by counterexample? *(p. 129)*
- How do you establish the falsity of a universal statement? *(p. 129)*

Proving a Universal Statement/Disproving an Existential Statement

- If a universal statement is defined over a small, finite domain, how do you use the method of exhaustion to prove that it is true? *(p. 130)*
- What is the method of generalizing from the generic particular? *(p. 130)*
- If you use the method of direct proof to prove a statement of the form "$\forall x$, if $P(x)$ then $Q(x)$", what do you suppose and what do you have to show? *(p. 131)*
- What are the guidelines for writing proofs of universal statements? *(p. 134)*
- What are some common mistakes people make when writing mathematical proofs? *(p. 135)*
- How do you disprove an existential statement? *(p. 138)*
- What is the method of proof by division into cases? *(p. 138)*
- If you use the method of proof by contradiction to prove a statement, what do you suppose and what do you have to show? *(p. 171)*
- If you use the method of proof by contraposition to prove a statement of the form "$\forall x$, if $P(x)$ then $Q(x)$", what do you suppose and what do you have to show? *(p. 175)*
- Are you able to use the various methods of proof and disproof to establish the truth or falsity of statements about odd and even integers *(p. 133)*, prime numbers *(p. 138)*, rational numbers *(pp. 143, 145, 146)*, divisibility of integers *(pp. 151-152)*, and the floor and ceiling of a real number *(pp. 166-168)*?

Some Important Theorems and Algorithms

- What is the theorem about divisibility by a prime number? *(p. 151)*
- What is the unique factorization theorem for the integers? (This theorem is also called the fundamental theorem of arithmetic.) *(p. 153)*
- What is the quotient-remainder theorem? Can you apply it to specific situations? *(p. 157)*
- What is the theorem about the irrationality of the square root of 2? Can you prove this theorem? *(p. 181)*

- What is the theorem about the infinitude of the prime numbers? Can you prove this theorem? *(p. 183)*
- What is the division algorithm ? *(p. 191)*
- What is the Euclidean algorithm? *(p. 192)*
- How do you use the Euclidean algorithm to compute the greatest common divisor of two positive integers? *(p. 195)*

Notation for Algorithms

- How is an assignment statement executed? *(p. 186)*
- How is an **if-then** statement executed? *(p. 187)*
- How is an **if-then-else** statement executed? *(p. 187)*
- How are the statements **do** and **end do** used in an algorithm? *(p. 187)*
- How is a **while** loop executed? *(p. 188)*
- How is a **for-next** loop executed? *(p. 189)*
- How do you construct a trace table for a segment of an algorithm? *(pp. 188-89, 191)*

Test Your Understanding: Chapter 3

Test yourself by filling in the blanks.

1. An integer is even if, and only if, it _____.

2. An integer is odd if, and only if, it _____.

3. An integer is prime if, and only if, _____.

4. An integer is composite if, and only if, _____.

5. If $n = 2k + 1$ for some integer k, then _____.

6. Given integers a and b, if there exists an integer k such that $b = ak$, then _____.

7. To find a counterexample for a statement of the form "$\forall x \in D$, if $P(x)$ then $Q(x)$" you find _____.

8. According to the method of generalizing from the generic particular, to prove that every element of a domain satisfies a certain property, you suppose that _____ and you show that _____.

9. According to the method of direct proof, to prove that a statement of the form "$\forall x$ in D, if $P(x)$ then $Q(x)$" is true, you suppose that _____ and you show that _____.

10. Proofs should always be written in _____ sentences, and each assertion made in a proof should be accompanied by a _____.

11. The fact that a universal statement is true in some instances does not imply that it is _____.

12. When writing a proof, it is a mistake to use the same letter to represent _____.

13. A real number is rational if, and only if, _____.

14. An integer a divides an integer b if, and only if, _____.

15. If a and b are integers, the notation $a \mid b$ stands for ____, and the notation a/b stands for ____.

16. According to the theorem about divisibility by a prime number, given any integer $n > 1$, there is a ____.

17. The unique factorization theorem (fundamental theorem of arithmetic) says that given any integer $n > 1$, n can be written as a ____ in a way that is unique, except possibly for the ____ in which the numbers are written.

18. The quotient-remainder theorem says that given any integer n and any positive integer d, there exists unique integers q and r such that ____.

19. If n is a nonnegative integer and d is a positive integer, then n div $d =$ ____ and n mod $d =$ ____ where ____.

20. The parity property says that any integer is either ____.

21. Suppose that at some point in a proof you know that one of the statements A_1 or A_2 or A_3 is true and you want to show that regardless of which statement happens to be true a certain conclusion C will follow. Then you need to show that ____ and ____ and ____.

22. Given any real number x, the floor of x is the unique integer n such that ____.

23. Given any real number x, the ceiling of x is the unique integer n such that ____.

24. To prove a statement by contradiction, you suppose that ____ and you show that ____.

25. To prove a statement of the form "$\forall x \in D$, if $P(x)$ then $Q(x)$" by contraposition, you suppose that ____ and you show that ____.

26. One way to prove that $\sqrt{2}$ is an irrational number is to assume that $\sqrt{2} = a/b$ for some integers a and b with no common factors greater than 1, use the lemma that says that if the square of an integer is even then ____, and eventually show that a and b ____.

27. One way to prove that there are infinitely many prime numbers is to assume that there are only finitely many prime numbers $p_1, p_2, \ldots, p_n$, construct the number ____, and then show that this number has to be divisible by a prime number that is greater than ____.

28. When an algorithm statement of the form $x := e$ is executed, ____.

29. Consider an algorithm statement of the following form.

    if (*condition*)
       then s_1
       else s_2

    When such a statement is executed, the truth or falsity of the *condition* is evaluated. If *condition* is true, ____. If *condition* is false, ____.

30. Consider an algorithm statement of the following form.

    while (*condition*)
       [statements that make up the body of the loop]
    end while

    When such a statement is executed, the truth or falsity of the *condition* is evaluated. If *condition* is true, ____. If *condition* is false, ____.

31. Consider an algorithm statement of the following form.

    for *variable* := *initial expression* **to** *final expression*
        [statements that make up the body of the loop]
    next *(same) variable*

    When such a statement is executed, *variable* is set equal to the value of the *initial expression*, and a check is made to determine whether the value of *variable* is less than or equal to the value of *final expression*. If so,_____. If not,_____.

32. Given a nonnegative integer a and a positive integer d, the division algorithm computes _____.

33. Given integers a and b, not both zero, $\gcd(a,b)$ is the integer d that satisfies the following two conditions: _____ and _____.

34. If r is a positive integer, then $\gcd(r,0) = $ _____.

35. If a and b are integers with $b \neq 0$ and if q and r are nonnegative integers such that $a = bq + r$, then $\gcd(a,b) = $ _____.

36. Given positive integers A and B with $A > B$, the Euclidean algorithm computes _____.

Answers

1. equals twice some integer
2. equals twice some integer plus 1
3. it is greater than 1, and if it is written as a product of positive integers, then one of the integers is 1
4. it is greater than 1, and it can be written as a product of positive integers neither of which is 1
5. n is an odd integer
6. a divides b (or $a \mid b$, or a is a factor of b; or a is a divisor of b; or b is divisible by a; or b is a multiple of a)
7. an element of D for which $P(x)$ is true and $Q(x)$ is false
8. you have a particular but arbitrarily chosen element of the domain
   that element satisfies the property
9. x is any *[particular but arbitrarily chosen]* element of D for which $P(x)$ is true
   $Q(x)$ is true
10. complete; reason that justifies the assertion
11. true in all instances
12. two different quantities
13. it can be written as a ratio of integers with a nonzero denominator
14. there is an integer, say k, such that $b = ak$
15. the sentence "a divides b"; the real number a divided by b (if $b \neq 0$)
16. prime number that divides n
17. product of prime numbers, order
18. $n = dq + r$ and $0 \leq r < d$
19. q; r; $n = dq + r$ and $0 \leq r < d$
20. even or odd
21. if A_1 is true then C is true; if A_2 is true then C is true; if A_3 is true then C is true
22. $n \leq x < n + 1$
23. $n - 1 < x \leq n$
24. the statement is false; this supposition leads to a contradiction

45. They are equal.

    Proof: Suppose m, n, and d are integers and $d \mid (m-n)$. By definition of divisibility, $m-n = dk$ for some integer k. Therefore, $m = n + dk$. Let $r = n \bmod d$. Then by definition of *mod*, $n = qd + r$ where q and r are integers and $0 \le r < d$. By substitution, $m = n + dk = (qd + r) + dk = d(q + k) + r$. Since $q + k$ is an integer and $0 \le r < d$, the integer quotient of the division of n by d is $q + k$ and the remainder is r. Hence $m \bmod d = r$ also, and so $n \bmod d = m \bmod d$.

48. *Proof:* Suppose m, d, and k are nonnegative integers and $d \ne 0$. Let $a = m \bmod d$. By definition of *mod*, $m = dq + a$ for some integer q and $0 \le a < d$. By substitution, $m + dk = dq + a + dk = d(q + k) + a$. Now $q + k$ is an integer because it is a sum of integers, and $0 \le a < d$. So by definition of *mod*, $(m + dk) \bmod d = a = m \bmod d$.

Section 3.5

6. If k is an integer, then $\lceil k \rceil = k$ because $k - 1 < k \le k$ and $k - 1$ and k are integers.

9. If the remainder obtained when n is divided by 36 is positive, an additional box beyond those containing exactly 36 units will be needed to hold the extra units. So since the ceiling notation rounds each number up to the nearest integer, the number of boxes required is $\lceil n/36 \rceil$. Also, because the ceiling of an integer is itself, if the number of units is a multiple of 36, the number of boxes required is $\lceil n/36 \rceil$ as well. Thus the ceiling notation is more appropriate for this problem because the answer is simply $\lceil n/36 \rceil$ regardless of the value of n. If the floor notation is used, the answer is more complicated: if $n/36$ is not an integer, it is $\lfloor n/36 \rfloor + 1$, but if n is an integer, it is $\lfloor n/36 \rfloor$.

18. *Counterexample:* Let $x = y = 1.5$. Then $\lceil x + y \rceil = \lceil 1.5 + 1.5 \rceil = \lceil 3 \rceil = 3$, whereas $\lceil x \rceil + \lceil y \rceil = \lceil 1.5 \rceil + \lceil 1.5 \rceil = 2 + 2 = 4$.

21. *Idea of Proof:* Given an odd integer n, you want to show that $\lceil \frac{n}{2} \rceil = \frac{n+1}{2}$ (*). The idea of the proof is first to express what it means for n to be odd and then to use that expression to rewrite both the left-hand side of equation (*) and the right-hand side of equation (*) and show that the resulting expressions are equal.

    Proof: Let n be any odd integer. *[We must show that $\lceil n/2 \rceil = (n+1)/2$.]* By definition of odd, $n = 2k + 1$ for some integer k. Substituting into the left-hand side of the equation to be proved gives

    $$\left\lceil \frac{n}{2} \right\rceil = \left\lceil \frac{2k+1}{2} \right\rceil = \left\lceil k + \frac{1}{2} \right\rceil = k + 1,$$

    where $\left\lceil k + \dfrac{1}{2} \right\rceil = k + 1$ by definition of ceiling because $k < k + 1/2 < k + 1$ and k is an integer. On the other hand, substituting into the right-hand side of the equation to be shown gives

    $$\frac{n+1}{2} = \frac{(2k+1)+1}{2} = \frac{2k+2}{2} = \frac{2(k+1)}{2} = k + 1$$

    also. Thus both the left- and right-hand sides of the equation to be proved equal $k + 1$, and so both are equal to each other. In other words, $\lceil n/2 \rceil = (n+1)/2$ *[as was to be shown]*.

24. *Idea of Proof:* Because of the nature of the given statement, the proof has to start with the supposition that m is any integer and x is any real number that is not an integer. Because the conclusion of the proof is an equation involving $\lfloor x \rfloor$ and $\lfloor m - x \rfloor$ and because $\lfloor x \rfloor$ is known to be an integer with a certain property, it is a good idea to give $\lfloor x \rfloor$ a specific name, say n. Then by definition of floor, $n \le x < n + 1$. But another part of the supposition is that x is not an integer. This means that the $\le$ sign can be replaced by a $<$ sign. Next, the fact that $\lfloor m - x \rfloor$

25. x is any *[particular but arbitrarily chosen]* element of D for which $Q(x)$ is false
    $P(x)$ is false

26. the integer is even; have a common factor greater than 1

27. $p_1 \cdot p_2 \cdots p_n + 1$; all the numbers $p_1, p_2, \ldots, p_n$

28. the expression e is evaluated (using the current values of all the variables in the expression), and this value is placed in the memory location corresponding to x (replacing any previous contents of the location)

29. statement s_1 is executed; statement s_2 is executed

30. all statements in the body of the loop are executed in order and then execution moves back to the beginning of the loop and the process repeats;
    execution passes to the next algorithm statement following the loop

31. the statements in the body of the loop are executed in order, *variable* is increased by 1, and execution returns to the top of the loop;
    execution passes to the next algorithm statement following the loop

32. integers q and r with the property that $n = dq + r$ and $0 \le r < d$

33. d divides a and d divides b; if c is a common divisor of both a and b, then $c \le d$

34. r

35. $\gcd(b, r)$

36. the greatest common divisor of A and B

Tips for Success with Proofs and Disproofs

Make sure your proofs are genuinely convincing. Express yourself carefully and completely – but concisely! Write in complete sentences, but don't use an unnecessary number of words.

Disproof by Counterexample

- To disprove a universal statement, give a counterexample.
- Write the word "Counterexample" at the beginning of a counterexample.
- Write counterexamples in complete sentences.
- Give values of the variables that you believe show the property is false.
- Include the computations that prove beyond any doubt that these values really do make the property false.

All Proofs

- Write the word "Proof" at the beginning of a proof.
- Write proofs in complete sentences.
- Start each sentence with a capital letter and finish with a period.

Direct Proof

- Begin each direct proof with the word "Suppose."
- In the "Suppose" sentence:
  - Introduce a variable or variables (indicating the general set they belong to - e.g., integers, real numbers etc.), and
  - Include the hypothesis that the variables satisfy.
- Identify the conclusion that you will need to show in order to complete the proof.
- Reason carefully from the "suppose" to the "conclusion to be shown."
- Include the little words (like "Then," "Thus," "So," "It follows that") that make your reasoning clear.
- Give a reason to support each assertion you make in your proof.

Proof by Contradiction

- Begin each proof by contradiction by writing "Suppose not. That is, suppose...," and continue this sentence by carefully writing the negation of the statement to be proved.
- After you have written the "suppose," you need to show that this supposition leads logically to a contradiction.
- Once you have derived a contradiction, you can conclude that the think you supposed is false. Since you supposed that the given statement was false, you now know that the given statement is true.

Proof by Contraposition

- Look to see if the statement to be proved is a universal conditional statement.
- If so, you can prove it by writing a direct proof of its contrapositive.

Chapter 4: Sequences and Mathematical Induction

The first section of this chapter introduces the notation for sequences, summations, products, and factorial. These chapters are intended to help you learn to recognize patterns so as to be able, for instance, to transform expanded versions of sums into summation notation, and to handle subscripts, particularly to change variables for summations and to distinguish index variables from variables that are constant with respect to a summation.

The second, third, and fourth sections of this chapter treat mathematical induction. The ordinary form is discussed in Sections 4.2 and 4.3 and the strong form in Section 4.4. Because of the importance of mathematical induction in discrete mathematics, a wide variety of examples is given to help you become comfortable with using the technique in many different situations.

The logic of ordinary mathematical induction is related to the logic discussed in Chapters 1 and 2. The main point is that the inductive step establishes the truth of a sequence of if-then statements. Together with the basis step, this sequence gives rise to a chain of inferences that lead to the desired conclusion. More formally:

Suppose
1. $P(1)$ is true; and
2. for all integers $k \geq 1$, if $P(k)$ is true then $P(k+1)$ is true.

The truth of statement (2) implies, according to the law of universal instantiation, that no matter what particular integer $k \geq 1$ is substituted in place of k, the statement "If $P(k)$ then $P(k+1)$" is true. The following argument, therefore, has true premises, and so by modus ponens it has a true conclusion:

	If $P(1)$ then $P(2)$.	by 2 and universal instantiation
	$P(1)$	by 1
$\therefore$	$P(2)$	by modus ponens

Similar reasoning gives the following chain of arguments, each of which has a true conclusion by modus ponens:

	If $P(2)$ then $P(3)$.
	$P(2)$
$\therefore$	$P(3)$
	If $P(3)$ then $P(4)$.
	$P(3)$
$\therefore$	$P(4)$
	If $P(4)$ then $P(5)$.
	$P(4)$
$\therefore$	$P(5)$
	And so forth.

Thus no matter how large a positive integer n is specified, the truth of $P(n)$ can be deduced as the final conclusion of a (possibly very long) chain of arguments continuing those shown above.

The concluding section of the chapter applies the technique of mathematical induction to proving algorithm correctness and is intended to be an introduction to the subject. It contains references to sources that treat the subject at length.

Section 4.1

6. $f_1 = \left\lfloor \dfrac{1}{4} \right\rfloor \cdot 4 = 0 \cdot 4 = 0, \quad f_2 = \left\lfloor \dfrac{2}{4} \right\rfloor \cdot 4 = 0 \cdot 4 = 0, \quad f_3 = \left\lfloor \dfrac{3}{4} \right\rfloor \cdot 4 = 0 \cdot 4 = 4,$

$f_4 = \left\lfloor \dfrac{4}{4} \right\rfloor \cdot 4 = 1 \cdot 4 = 4$

9.

$$
\begin{aligned}
h_1 &= 1 \cdot \lfloor \log_2 1 \rfloor &= 1 \cdot 0 \\
h_2 &= 2 \cdot \lfloor \log_2 2 \rfloor &= 2 \cdot 1 \\
h_3 &= 3 \cdot \lfloor \log_2 3 \rfloor &= 3 \cdot 1 \\
h_4 &= 4 \cdot \lfloor \log_2 4 \rfloor &= 4 \cdot 2 \\
h_5 &= 5 \cdot \lfloor \log_2 5 \rfloor &= 5 \cdot 2 \\
h_6 &= 6 \cdot \lfloor \log_2 6 \rfloor &= 6 \cdot 2 \\
h_7 &= 7 \cdot \lfloor \log_2 7 \rfloor &= 7 \cdot 2 \\
h_8 &= 8 \cdot \lfloor \log_2 8 \rfloor &= 8 \cdot 3 \\
h_9 &= 9 \cdot \lfloor \log_2 9 \rfloor &= 9 \cdot 3 \\
h_{10} &= 10 \cdot \lfloor \log_2 10 \rfloor &= 10 \cdot 3 \\
h_{11} &= 11 \cdot \lfloor \log_2 11 \rfloor &= 11 \cdot 3 \\
h_{12} &= 12 \cdot \lfloor \log_2 12 \rfloor &= 12 \cdot 3 \\
h_{13} &= 13 \cdot \lfloor \log_2 13 \rfloor &= 13 \cdot 3 \\
h_{14} &= 14 \cdot \lfloor \log_2 14 \rfloor &= 14 \cdot 3 \\
h_{15} &= 15 \cdot \lfloor \log_2 15 \rfloor &= 15 \cdot 3
\end{aligned}
$$

When n is an integral power of 2, h_n is n times the exponent of that power. For instance, $8 = 2^3$ and $h_8 = 8 \cdot 3$. If m and n are integers and $2^m \leq 2^n < 2^{m+1}$, then $h_n = n \cdot m$.

15. One answer is $a_n = (-1)^n \left(\dfrac{n-1}{n} \right)$ for all integers $n \geq 1$. Another possible answer is $a_n = (-1)^{n+1} \left(\dfrac{n}{n+1} \right)$ for all integers $n \geq 0$. Additional correct answers are possible.

18. *e.* $\prod_{k=2}^{2} a_k = a_2 = -2$

21. $\displaystyle\sum_{m=0}^{3} \dfrac{1}{2^m} = \dfrac{1}{2^0} + \dfrac{1}{2^1} + \dfrac{1}{2^2} + \dfrac{1}{2^3} = 1 + \dfrac{1}{2} + \dfrac{1}{4} + \dfrac{1}{8} = \dfrac{15}{8}$

24. $\displaystyle\sum_{j=0}^{0} (j+1) \cdot 2^j = (0+1) \cdot 2^0 = 1 \cdot 1 = 1$

30. $1 \cdot 2 + 2 \cdot 3 + 3 \cdot 4 + \cdots + n \cdot (n+1)$

33. One possible answer is $\displaystyle\sum_{k=1}^{5} (-1)^{k+1}(k^3 - 1)$. Other correct answers are possible.

39. One possible answer is $\displaystyle\sum_{k=1}^{n} \dfrac{k}{(k+1)!}$. Other correct answers are possible.

51. *b. Proof:* Let n and k be integers with $n \geq 2$ and $2 \leq k \leq n$. Now $n!$ is the product of all the integers from 1 to n, and so since $n \geq 2$ and $2 \leq k \leq n$, k is a factor of $n!$. That is $n! = kr$ for some integer r. By substitution $n! + k = kr + k = k(r + 1)$. But $r + 1$ is an integer because r is. By definition of divisibility, therefore, $n! + k$ is divisible by k.

c. Yes. If m is any integer that is greater than or equal to 2, then none of the terms of the following sequence of integers is prime: $m! + 2, m! + 3, m! + 4, \ldots, m! + m$. The reason is that each has the form $m! + k$ for an integer k with $2 \leq k \leq m$, and for each such k, by part (b) $m! + k$ is divisible by k.

57. When $i = n$, $j = n - 1$. When $i = 2n$, $j = 2n - 1$. Since $j = i - 1$, then $i = j + 1$. So $\dfrac{n - i + 1}{n + i} = \dfrac{n - (j+1) + 1}{n + j + 1} = \dfrac{n - j}{n + j + 1}$. Therefore, $\displaystyle\prod_{i=n}^{2n} \left(\dfrac{n - i + 1}{n + i} \right) = \prod_{j=n-1}^{2n-1} \left(\dfrac{n - j}{n + j + 1} \right)$.

60. By Theorem 4.1.1, $\left(\prod_{k=1}^{n}(\frac{k}{k+1})\right)\left(\prod_{k=1}^{n}(\frac{k+1}{k+2})\right) = \prod_{k=1}^{n}(\frac{k}{k+1})(\frac{k+1}{k+2}) = \prod_{k=1}^{n}(\frac{k}{k+2})$.

69. Let a nonnegative integer a be given. Divide a by 16 using the quotient-remainder theorem to obtain a quotient $q[0]$ and a remainder $r[0]$. If the quotient is nonzero, divide by 16 again to obtain a quotient $q[1]$ and a remainder $r[1]$. Continue this process until a quotient of 0 is obtained. The remainders calculated in this way are the hexadecimal digits of a:

$$a_{10} = (r[k]r[k-1]\ldots r[2]r[1]r[0])_{16}.$$

72.

$$\begin{array}{r|l}
 & 0 \qquad \text{R. } 8 = 8_{16} \\
16 & 8 \qquad \text{R. } 15 = F_{16} \\
16 & 143 \quad \text{R. } 13 = D_{16} \\
16 & 2301 \\
\end{array}$$
$\qquad\qquad\qquad$ Hence $2301_{10} = 8FD_{16}$.

Section 4.2

9. *Proof (by mathematical induction)*: Let the property $P(n)$ be the equation

$$4^3 + 4^4 + 4^5 + \cdots + 4^n = \frac{4(4^n - 16)}{3}.$$

Show that the property is true for $n = 3$: The property is true for $n = 3$ because for $n = 3$ the left-hand side is $4^3 = 64$ and the right-hand side is $\frac{4(4^3 - 16)}{3} = \frac{4(64 - 16)}{3} = \frac{4 \cdot 48}{3} = 64$ also.

Show that for all integers $k \geq 1$, if the property is true for $n = k$ then it is true for $n = k + 1$: Suppose $4^3 + 4^4 + 4^5 + \cdots + 4^k = \frac{4(4^k - 16)}{3}$ for some integer $k \geq 3$. *[This is the inductive hypothesis.]* We must show that $4^3 + 4^4 + 4^5 + \cdots + 4^{k+1} = \frac{4(4^{k+1} - 16)}{3}$. But the left-hand side of this equation is

$$
\begin{aligned}
4^3 + 4^4 + 4^5 + \cdots + 4^{k+1} \quad &= \quad 4^3 + 4^4 + 4^5 + \cdots + 4^k + 4^{k+1} \\
& \qquad \text{by making the next-to-last term explicit} \\
&= \quad \frac{4(4^k - 16)}{3} + 4^{k+1} \\
& \qquad \text{by inductive hypothesis} \\
&= \quad \frac{4^{k+1} - 64}{3} + \frac{3 \cdot 4^{k+1}}{3} \\
& \qquad \text{by creating a common denominator} \\
&= \quad \frac{4 \cdot 4^{k+1} - 64}{3} \\
& \qquad \text{by adding the fractions} \\
&= \quad \frac{4(4^{k+1} - 16)}{3} \\
& \qquad \text{by factoring out the 4,}
\end{aligned}
$$

and this is the right-hand side of the equation *[as was to be shown]*.

12. *Proof (by mathematical induction):* Let the property $P(n)$ be the equation

$$\frac{1}{1 \cdot 2} + \frac{1}{2 \cdot 3} + \cdots + \frac{1}{n(n+1)} = \frac{n}{n+1}.$$

Show that the property is true for $n = 1$: The property is true for $n = 1$ because the left-hand side equals $\frac{1}{1 \cdot 2} = \frac{1}{2}$ and the right-hand side equals $\frac{1}{1+1} = \frac{1}{2}$ also.

Show that for all integers $k \geq 1$, ***if the property is true for*** $n = k$ ***then it is true for*** $n = k + 1$: Suppose $\frac{1}{1 \cdot 2} + \frac{1}{2 \cdot 3} + \cdots + \frac{1}{k(k+1)} = \frac{k}{k+1}$ for some integer $k \geq 1$. *[This is the inductive hypothesis.]* We must show that $\frac{1}{1 \cdot 2} + \frac{1}{2 \cdot 3} + \cdots + \frac{1}{(k+1)((k+1)+1)} = \frac{k+1}{(k+1)+1}$, or, equivalently, $\frac{1}{1 \cdot 2} + \frac{1}{2 \cdot 3} + \cdots + \frac{1}{(k+1)(k+2)} = \frac{k+1}{k+2}$. But the left-hand side of this equation is

$$\frac{1}{1 \cdot 2} + \frac{1}{2 \cdot 3} + \cdots + \frac{1}{(k+1)(k+2)}$$

$$= \frac{1}{1 \cdot 2} + \frac{1}{2 \cdot 3} + \cdots + \frac{1}{k(k+1)} + \frac{1}{(k+1)(k+2)}$$

by making the next-to-last term explicit

$$= \frac{k}{k+1} + \frac{1}{(k+1)(k+2)}$$

by inductive hypothesis

$$= \frac{k(k+2)}{(k+1)(k+2)} + \frac{1}{(k+1)(k+2)}$$

by creating a common denominator

$$= \frac{k^2 + 2k + 1}{(k+1)(k+2)}$$

by adding the fractions

$$= \frac{(k+1)^2}{(k+1)(k+2)}$$

because $k^2 + 2k + 1 = (k+1)^2$

$$= \frac{k+1}{k+2}$$

by cancelling $(k+1)$ from numerator and denominator,

and this is the right-hand side of the equation *[as was to be shown]*.

15. *Proof (by mathematical induction):* Let the property $P(n)$ be the equation

$$\sum_{i=1}^{n} i(i!) = (n+1)! - 1.$$

Show that the property is true for $n = 1$: The property holds for $n = 1$ because $\sum_{i=1}^{1} i(i!) = 1 \cdot (1!) = 1$ and $(1+1)! - 1 = 2! - 1 = 2 - 1 = 1$ also.

Show that for all integers $k \geq 1$, ***if the property is true for*** $n = k$ ***then it is true for*** $n = k + 1$: Suppose $\sum_{i=1}^{k} i(i!) = (k+1)! - 1$ for some integer $k \geq 1$. *[This is the inductive hypothesis.]* We must show that $\sum_{i=1}^{k+1} i(i!) = ((k+1)+1)! - 1$, or, equivalently, we must show that $\sum_{i=1}^{k+1} i(i!) = (k+2)! - 1$. But the left-hand side of the equation is

$$\begin{aligned} \sum_{i=1}^{k+1} i(i!) &= \sum_{i=1}^{k} i(i!) + (k+1)((k+1)!) && \text{by writing the } (k+1)\text{st term separately} \\ &= [(k+1)! - 1] + (k+1)((k+1)!) && \text{by inductive hypothesis} \\ &= ((k+1)!)(1 + (k+1)) - 1 && \text{by combining the terms with} \\ &&& \text{the common factor } (k+1)! \\ &= (k+1)!(k+2) - 1 \\ &= (k+2)! - 1 && \text{by algebra,} \end{aligned}$$

and this is the right-hand side of the equation *[as was to be shown]*.

18. *Proof (by mathematical induction):* Let the property $P(n)$ be the equation

$$\sin x + \sin 3x + \cdots + \sin(2n-1)x = \frac{1 - \cos 2nx}{2\sin x}.$$

Show that the property is true for $n = 1$: The property holds for $n = 1$ because the left-hand side equals $\sin x$, and the right-hand side equals $\dfrac{1 - \cos 2x}{2\sin x} = \dfrac{1 - \cos^2 + \sin^2 x}{2\sin x} = \dfrac{2\sin^2 x}{2\sin x} = \sin x$.

Show that for all integers $k \geq 1$, if the property is true for $n = k$ then it is true for $n = k + 1$: Suppose $\sin x + \sin 3x + \cdots + \sin(2k-1)x = \dfrac{1 - \cos 2kx}{2\sin x}$ for some integer $k \geq 1$. *[This is the inductive hypothesis.]* We must show that $\sin x + \sin 3x + \cdots + \sin(2(k+1)-1)x = \dfrac{1 - \cos 2(k+1)x}{2\sin x}$, or, equivalently, $\sin x + \sin 3x + \cdots + \sin(2k+1)x = \dfrac{1 - \cos 2(k+1)x}{2\sin x}$. But when the next-to-last term of the the left-hand side of this equation is made explicit, the left-hand side becomes

$\sin x + \sin 3x + \cdots + \sin(2k-1)x + \sin(2k+1)x$

$$
\begin{aligned}
&= \frac{1 - \cos 2kx}{2\sin x} + \sin(2k+1)x && \text{by inductive hypothesis}\\[4pt]
&= \frac{1 - \cos 2kx}{2\sin x} + \frac{2\sin x \sin(2kx+x)}{2\sin x} && \text{by creating a common denominator}\\[4pt]
&= \frac{1 - \cos 2kx + 2\sin x \sin(2kx+x)}{2\sin x} && \text{by adding fractions}\\[4pt]
&= \frac{1 - \cos 2kx + 2\sin x[\sin(2kx)\cos x + \cos(2kx)\sin x]}{2\sin x} && \text{by the addition formula for sine}\\[4pt]
&= \frac{1 - \cos 2kx + 2\sin x \sin(2kx)\cos x + 2\sin^2 x \cos(2kx)}{2\sin x} && \text{by multiplying out}\\[4pt]
&= \frac{1 + \cos 2kx(2\sin^2 x - 1) + 2\sin x \cos x \sin(2kx)}{2\sin x} && \text{by combining like terms}\\[4pt]
&= \frac{1 + \cos 2kx(-\cos 2x) + \sin 2x \sin(2kx)}{2\sin x} && \text{by the formulas for } \cos 2x \text{ and } \sin 2x\\[4pt]
&= \frac{1 - (\cos 2kx \cos 2x - \sin 2x \sin(2kx))}{2\sin x} && \text{by factoring out } -1\\[4pt]
&= \frac{1 - \cos(2kx + 2x)}{2\sin x} && \text{by the addition formula for cosine}\\[4pt]
&= \frac{1 - \cos(2(k+1)x)}{2\sin x} && \text{by factoring out 2x,}
\end{aligned}
$$

and this is the right-hand side of the equation *[as was to be shown]*.

30. $ar^m + ar^{m+1} + ar^{m+2} + \cdots + ar^{m+n} = ar^m(1 + r + r^2 + \cdots + r^n) = ar^m\left(\dfrac{r^{n+1} - 1}{r - 1}\right)$ by Theorem 4.2.3

33. *Proof:* Suppose m and n are any positive integers such that m is odd. By definition of odd, $m = 2q + 1$ for some integer k, and so, by Theorems 4.1.1 and 4.2.2,

$$\sum_{k=0}^{m-1}(n+k) = \sum_{k=0}^{(2q+1)-1}(n+k) = \sum_{k=0}^{2q}(n+k) = \sum_{k=0}^{2q}n + \sum_{k=0}^{2q}k = (2q+1)n + \sum_{k=1}^{2q}k$$

$$= (2q+1)n + \frac{2q(2q+1)}{2} = (2q+1)n + q(2q+1) = (2q+1)(n+q) = m(n+q).$$

But $n+q$ is an integer because it is a sum of integers. Hence, by definition of divisibility, $\sum_{k=0}^{m-1}(n+k)$ is divisible by m.

Note: If m is even, the property is no longer true. For example, if $n=1$ and $m=2$, then $\sum_{k=0}^{m-1}(n+k) = \sum_{k=0}^{2-1}(1+k) = 1+2 = 3$, and 3 is not divisible by 2.

Section 4.3

9. *Proof (by mathematical induction)*: Let the property $P(n)$ be the sentence "$7^n - 1$ is divisible by 6."

Show that the property is true for $n=0$: The property is true for $n=0$ because $7^0 - 1 = 1 - 1 = 0$ and 0 is divisible by 6 (since $0 = 0 \cdot 6$).

Show that for all integers $k \geq 0$, if the property is true for $n=k$ then it is true for $n=k+1$: Suppose $7^k - 1$ is divisible by 6 for some integer $k \geq 0$. *[This is the inductive hypothesis.]* We must show that $7^{k+1} - 1$ is divisible by 6. By definition of divisibility, the inductive hypothesis is equivalent to the statement $7^k - 1 = 6r$ for some integer r. Then by the laws of algebra, $7^{k+1} - 1 = 7 \cdot 7^k - 1 = (6+1)7^k - 1 = 6 \cdot 7^k + (7^k - 1) = 6 \cdot 7^k + 6r$, where the last equality holds by inductive hypothesis. Thus, by factoring out the 6 from the extreme right-hand side and by equating the extreme left-hand and extreme right-hand sides, , we have $7^{k+1} - 1 = 6(7^k + r)$, which is divisible by 6 because $7^k + r$ is an integer (since products and sums of integers are integers). Therefore, $7^{k+1} - 1$ is divisible by 6 *[as was to be shown]*.

12. *Proof (by mathematical induction)*: Let the property $P(n)$ be the sentence "$7^n - 2^n$ is divisible by 5."

Show that the property is true for $n=1$: The property is true for $n=1$ because $7^1 - 2^1 = 7 - 2 = 5$ and 5 is divisible by 5 (since $5 = 5 \cdot 1$).

Show that for all integers $k \geq 1$, if the property is true for $n=k$ then it is true for $n=k+1$: Suppose $7^k - 2^k$ is divisible by 5 for some integer $k \geq 0$. *[This is the inductive hypothesis.]* We must show that $7^{k+1} - 2^{k+1}$ is divisible by 5. By definition of divisibility, the inductive hypothesis is equivalent to the statement $7^k - 2^k = 5r$ for some integer r. Then by the laws of algebra, $7^{k+1} - 2^{k+1} = 7 \cdot 7^k - 2 \cdot 2^k = (5+2) \cdot 7^k - 2 \cdot 2^k = 5 \cdot 7^k + 2 \cdot 7^k - 2 \cdot 2^k = 5 \cdot 7^k + 2(7^k - 2^k) = 5 \cdot 7^k + 2 \cdot 5r$, where the last equality holds by inductive hypothesis. Thus, by factoring out the 5 from the extreme right-hand side and by equating the extreme left-hand and extreme right-hand sides, we have $7^{k+1} - 2^{k+1} = 5(7^k + 2r)$, which is divisible by 5 because $7^k + 2r$ is an integer (since products and sums of integers are integers). Therefore, $7^{k+1} - 2^{k+1}$ is divisible by 5 *[as was to be shown]*.

15. *Proof (by mathematical induction)*: Let the property $P(n)$ be the sentence "$n(n^2+5)$ is divisible by 6."

Show that the property is true for $n=1$: The property is true for $n=1$ because $1(1^2 + 5) = 6$ and 6 is divisible by 6.

Show that for all integers $k \geq 1$, if the property is true for $n=k$ then it is true for $n=k+1$: Suppose $k(k^2+5)$ is divisible by 6 for some integer $k \geq 1$. *[This is the inductive hypothesis.]* We must show that $(k+1)((k+1)^2+5)$ is divisible by 6. By definition of divisibility $k(k^2+5) = 6r$ for some integer r. Then by the laws of algebra, $(k+1)((k+1)^2+5) = (k+1)(k^2+2k+1+5) = k(k^2+5) + (k(2k+1)+k^2+2k+1+5) = k(k^2+5) + (3k^2+3k+6) = 6r + 3(k^2+k) + 6$, where the last equality holds by inductive hypothesis. Now $k(k+1)$ is a product of two consecutive integers. By Theorem 3.4.2 one of these is even, and so *[by Section 3.1, exercise 42 or Example 3.2.3]* the product $k(k+1)$ is even. Hence $k(k+1) = 2s$ for some integer s. Thus $6r + 3(k^2 + k) + 6 = 6r + 3(2s) + 6 = 6(r + s + 1)$. By substitution, then, $(k+1)((k+1)^2 + 5) = 6(r + s + 1)$, which is divisible by 6 because $r + s + 1$ is an integer. Therefore, $(k+1)((k+1)^2 + 5)$ is divisible by 6 *[as was to be shown]*.

18. *Proof (by mathematical induction)*: Let the property $P(n)$ be the inequality $5^n + 9 < 6^n$.

    Show that the property is true for $n = 2$: The property is true for $n = 2$ because the left-hand side is $5^2 + 9 = 25 + 9 = 34$ and the right-hand side is $6^2 = 36$, and $34 < 36$.

    Show that for all integers $k \geq 2$, if the property is true for $n = k$ then it is true for $n = k + 1$: Suppose $5^k + 9 < 6^k$ for some integer $k \geq 2$. *[This is the inductive hypothesis.]* We must show that $5^{k+1} + 9 < 6^{k+1}$. Multiplying both sides of the inequality in the inductive hypothesis by 5 gives $5(5^k + 9) < 5 \cdot 6^k$. Note that $5(5^k + 9) = 5^{k+1} + 45$, $5 \cdot 6^k < 6^{k+1}$, and $5^{k+1} + 9 < 5^{k+1} + 45$. Putting these together gives $5^{k+1} + 9 < 5^{k+1} + 45 < 5 \cdot 6^k < 6^{k+1}$, and so, by transitivity of order, $5^{k+1} + 9 < 6^{k+1}$ *[as was to be shown]*.

21. *Proof (by mathematical induction)*: Let the property $P(n)$ be the inequality

$$\sqrt{n} < \frac{1}{\sqrt{1}} + \frac{1}{\sqrt{2}} + \frac{1}{\sqrt{3}} + \cdots + \frac{1}{\sqrt{n}}.$$

    Show that the property is true for $n = 2$: To show that the property is true for $n = 2$ we must show that $\sqrt{2} < \frac{1}{\sqrt{1}} + \frac{1}{\sqrt{2}}$. But this inequality is true if, and only if, $2 < \sqrt{2} + 1$ (by multiplying/dividing both sides by $\sqrt{2}$). And this is true if, and only if, $1 < \sqrt{2}$ (by subtracting/adding 1 on both sides). But $1 < \sqrt{2}$, and so the inequality holds for $n = 2$.

    Show that for all integers $k \geq 2$, if the property is true for $n = k$ then it is true for $n = k + 1$: Suppose $\sqrt{k} < \frac{1}{\sqrt{1}} + \frac{1}{\sqrt{2}} + \frac{1}{\sqrt{3}} + \cdots + \frac{1}{\sqrt{k}}$ for some integer $k \geq 2$. *[This is the inductive hypothesis.]* We must show that $\sqrt{k+1} < \frac{1}{\sqrt{1}} + \frac{1}{\sqrt{2}} + \frac{1}{\sqrt{3}} + \cdots + \frac{1}{\sqrt{k+1}}$. But for each integer $k \geq 2$, $\sqrt{k} < \sqrt{k+1}$ (*), and so (by multiplying both sides by $\sqrt{k}$) $k < \sqrt{k} \cdot \sqrt{k+1}$. Adding 1 to both sides gives $k + 1 < \sqrt{k} \cdot \sqrt{k+1} + 1$, and dividing both sides by $\sqrt{k+1}$ gives $\sqrt{k+1} < \sqrt{k} + \frac{1}{\sqrt{k+1}}$. By substitution from the inductive hypothesis, then, $\sqrt{k+1} < \frac{1}{\sqrt{1}} + \frac{1}{\sqrt{2}} + \frac{1}{\sqrt{3}} + \cdots + \frac{1}{\sqrt{k}} + \frac{1}{\sqrt{k+1}}$ *[as was to be shown]*.

    (*) *Note*: Strictly speaking, the reason for this claim is that $k < k + 1$ and for all positive real numbers a and b, if $a < b$, then $\sqrt{a} < \sqrt{b}$.

27. *Proof by mathematical induction*: According to the definition of $d_1, d_2, d_3, \ldots$, $d_1 = 2$ and $d_k = \frac{d_{k-1}}{k}$ for all integers $k \geq 2$. Let the property $P(n)$ be the equation $d_n = \frac{2}{n!}$.

    Show that the property is true for $n = 1$: We must show that $d_1 = \frac{2}{1!}$. But $\frac{2}{1!} = 2$ and $d_1 = 2$ by definition of $d_1, d_2, d_3, \ldots$. So the property holds for $n = 1$.

    Show that for all integers $k \geq 1$, if the property is true for $n = k$ then it is true for $n = k + 1$: Suppose that for some integer $k \geq 1$, $d_k = \frac{2}{k!}$. *[This is the inductive hypothesis.]*

We must show that $d_{k+1} = \dfrac{2}{(k+1)!}$. But

$$
\begin{aligned}
d_{k+1} &= \frac{d_k}{k+1} && \text{by definition of } d_1, d_2, d_3, \ldots \\[2mm]
&= \frac{\frac{2}{k!}}{k+1} && \text{by inductive hypothesis} \\[2mm]
&= \frac{2}{(k+1)k!} \\[2mm]
&= \frac{2}{(k+1)!} && \text{by the algebra of fractions.}
\end{aligned}
$$

[This is what was to be shown].

33. Let $P(n)$ be the property that "in any round-robin tournament involving n teams, it is possible to label the teams $T_1, T_2, T_3, \ldots, T_n$ so that T_i beats T_{i+1} for all $i = 1, 2, 3, \ldots, n-1$." We will show by mathematical induction that this property is true for all integers $n \geq 2$.

Show that the property is true for $n = 2$: Consider any round-robin tournament involving two teams. By definition of round-robin tournament, these teams play each other exactly once. Let T_1 be the winner and T_2 the loser of this game. Then T_1 beats T_2, and so the labeling is as required.

Show that for all integers $k \geq 2$, if the property is true for $n = k$ then it is true for $n = k+1$: Let k be an integer with $k \geq 2$ and suppose that in any round-robin tournament involving k teams it is possible to label the teams in the way described. *[This is the inductive hypothesis.]* We must show that in any round-robin tournament involving $k+1$ teams it is possible to label the teams in the way described.

Consider any round-robin tournament with $k+1$ teams. Pick one and call it T'. Temporarily remove T' and consider the remaining k teams. Since each of these teams plays each other team exactly once, the games played by these k teams form a round-robin tournament. It follows by inductive hypothesis that these k teams may be labeled $T_1, T_2, T_3, \ldots, T_k$ where T_i beats T_{i+1} for all $i = 1, 2, 3, \ldots, k-1$.

Case 1 (T' beats T_1): In this case, relabel each T_i to be T_{i+1}, and let $T_1 = T'$. Then T_1 beats the newly labeled T_2 (because T' beats the old T_1), and T_i beats T_{i+1} for all $i = 2, 3, \ldots, k$ (by inductive hypothesis).

Case 2 (T' loses to $T_1, , T_2, T_3, \ldots, T_m$ and beats T_{m+1} where $1 \leq m \leq k-1$): In this case, relabel teams $T_{m+1}, T_{m+2}, \ldots, T_k$ to be $T_{m+2}, T_{m+3}, \ldots, T_{k+1}$ and let $T_{m+1} = T'$. Then for each i with $1 \leq i \leq m-1$, T_i beats T_{i+1} (by inductive hypothesis), T_m beats T_{m+1} (because T_m beats T'), T_{m+1} beats T_{m+2} (because T' beats the old T_{m+1}), and for each i with $m+2 \leq i \leq k$, T_i beats T_{i+1} (by inductive hypothesis).

Case 3 (T' loses to T_i for all $i = 1, 2, \ldots, k$): In this case, let $T_{k+1} = T'$. Then for all $i = 1, 2, \ldots, k-1$, T_i beats T_{i+1} (by inductive hypothesis) and T_k beats T_{k+1} (because T_k beats T').

Thus in all three cases the teams may be relabeled in the way specified *[as was to be shown]*.

Section 4.4

3. *Proof:* Let the property $P(n)$ be the sentence "c_n is even." We prove by strong mathematical induction that this property is true for all integers $n \geq 0$.

Show that the property is true for $n = 0$, $n = 1$, and $n = 2$: $c_0 = 2$, $c_1 = 2$, and $c_2 = 6$ and 2, 2, and 6 are all even. So the property is true for $n = 0, 1,$ and 2.

Show that if $k > 2$ *and the property is true for all integers* i *with* $0 \leq i < k$, *then it is true for* $n = k$: Let $k > 2$ be an integer, and suppose c_i is even for all integers i with $1 \leq i < k$. *[This is the inductive hypothesis.]* We must show that c_k is even. But by definition of $c_0, c_1, c_2, \ldots, c_k = 3c_{k-3}$. Since $k > 2$, $0 \leq k - 3 < k$, and so, by inductive hypothesis, c_{k-3} is even. But the product of an even integer with any integer is even *[exercise 42, Section 3.1 or Example 3.2.3]*, and so $3c_{k-3}$, which equals c_k, is also even *[as was to be shown]*.

6. *Proof*: Let the property $P(n)$ be the inequality $f_n \leq n$. We prove by strong mathematical induction that this property is true for all integers $n \geq 1$.

   Show that the property is true for $n = 1$: $f_1 = 1$ and $1 \leq 1$. So the property is true for $n = 1$.

   Show that if $k > 1$ *and the property is true for all integers* i *with* $1 \leq i < k$, *then it is true for* $n = k$: Let $k > 1$ be an integer, and suppose $f_i \leq i$ for all integers i with $1 \leq i < k$. *[This is the inductive hypothesis.]* We must show that $f_k \leq k$. But by definition of $f_1, f_2, f_3, \ldots, f_k = 2 \cdot f_{\lfloor k/2 \rfloor}$. Since $k > 1$, $1 \leq \lfloor k/2 \rfloor < k$, and so, by inductive hypothesis, $f_{\lfloor k/2 \rfloor} \leq \lfloor k/2 \rfloor$. Thus

$$f_k = 2 \cdot f_{\lfloor k/2 \rfloor} \leq 2 \cdot \lfloor k/2 \rfloor = \begin{cases} 2 \cdot ((k-1)/2) & \text{if } k \text{ is odd} \\ 2 \cdot (k/2) & \text{if } k \text{ is even} \end{cases} = \begin{cases} k - 1 & \text{if } k \text{ is odd} \\ k & \text{if } k \text{ is even} \end{cases} \leq k,$$

and so $f_k \leq k$ *[as was to be shown]*.

9. *Proof*: Let the property $P(n)$ be the inequality $a_n \leq \left(\dfrac{7}{4}\right)^n$. We prove by strong mathematical induction that this property is true for all integers $n \geq 1$.

   Show that the property is true for $n = 1$ *and* $n = 2$: By definition of $a_1, a_2, a_3, \ldots$, $a_1 = 1$ and $a_2 = 3$. But $1 < \dfrac{7}{4}$ and $\left(\dfrac{7}{4}\right)^2 = \dfrac{49}{16} = 3\dfrac{1}{16} > 3$. So $a_1 \leq \dfrac{7}{4}$ and $a_2 \leq \left(\dfrac{7}{4}\right)^2$, and thus the property is true for $n = 1$ and $n = 2$.

   Show that if $k > 2$ *and the property is true for all integers* i *with* $1 \leq i < k$, *then it is true for* $n = k$: Let $k > 2$ be an integer, and suppose $a_i \leq \left(\dfrac{7}{4}\right)^i$ for all integers i with $1 \leq i < k$. *[This is the inductive hypothesis.]* We must show that $a_k \leq \left(\dfrac{7}{4}\right)^k$. But by definition of $a_1, a_2, a_3, \ldots$, $a_k = a_{k-1} + a_{k-2}$. Since $k > 2$, $1 \leq k - 2 < k - 1 < k$, and so by inductive hypothesis, $a_{k-1} \leq \left(\dfrac{7}{4}\right)^{k-1}$ and $a_{k-2} \leq \left(\dfrac{7}{4}\right)^{k-2}$. Adding the inequalities and using the laws of basic algebra gives

$$a_{k-1} + a_{k-2} \leq \left(\frac{7}{4}\right)^{k-1} + \left(\frac{7}{4}\right)^{k-2} = \left(\frac{7}{4}\right)^{k-2}\left(\frac{7}{4} + 1\right) = \left(\frac{7}{4}\right)^{k-2}\left(\frac{11}{4}\right)$$

$$= \left(\frac{7}{4}\right)^{k-2}\left(\frac{44}{16}\right) < \left(\frac{7}{4}\right)^{k-2}\left(\frac{49}{16}\right) = \left(\frac{7}{4}\right)^{k-2}\left(\frac{7}{4}\right)^2 = \left(\frac{7}{4}\right)^k.$$

So $a_{k-1} + a_{k-2} \leq \left(\dfrac{7}{4}\right)^k$ *[as was to be shown]*.

24. No.

   Counterexample 1: Let $P(n)$ be "$n \neq 4$." Then $P(0)$, $P(1)$, and $P(2)$ are all true (because $0 \neq 4$, $1 \neq 4$, and $2 \neq 4$,), and for all integers $k \geq 0$, if $P(k)$ is true then $P(3k)$ is true (because if $k \neq 4$ then $3k \neq 4$). But when $n = 4$, the statement $n \neq 4$ is false. So it is not the case that $P(n)$ is true for all integers $n \geq 0$

Counterexample 2: Let $P(n)$ be "$n + 24$ is composite." Then $P(0)$, $P(1)$, and $P(2)$ are all true (because 24, 25, and 26 are composite), and for all integers $k \geq 0$, if $P(k)$ is true then $P(3k)$ is true (because if $k + 24$ is composite then $3k + 24$ is also composite — in fact, $3k + 24$ is composite for *any* integer $k \geq 0$). But it is false that $n + 24$ is composite for all integers $n \geq 0$ (because, for instance, $5 + 24 = 29$ is prime).

Counterexample 3: Let $P(n)$ be "$n = 0$, or $n = 1$, or $n = 2$, or $n = 3r$ for some integer r." Then $P(0)$, $P(1)$, and $P(2)$ are all true (because an *or* statement is true if any component is true), and for all integers $k \geq 0$, if $P(k)$ is true then $P(3k)$ is true (also by definition of the truth value of an *or* statement and by definition of $P(n)$). But $P(n)$ is not true for all integers $n \geq 0$ (because, for instance, $P(4)$ is not true).

(There are many other counterexamples besides these two.)

27. *Proof*: Consider the property "n can be written in the form

$$n = c_r \cdot 3^r + c_{r-1}3^{r-1} + \cdots + c_2 \cdot 3^2 + c_1 \cdot 3 + c_0$$

where r is a nonnegative integer, $c_r = 1$ or 2 and $c_j = 0$, 1, or 2 for all $j = 0, 1, 2, \ldots, r - 1$."
We will show that the property is true for all integers $n \geq 1$.

Show that the property is true for $n = 1$: Observe that $1 = c_r 3^r$ for $r = 0$ and $c_r = 1$. Thus 1 can be written in the required form.

Show that if $k > 1$ and the property is true for all integers i with $1 \leq i < k$, then it is true for $n = k$: Let k be an integer with $k > 1$, and suppose that for all integers i with $1 \leq i < k$, i can be written in the required form:

$$i = c_r \cdot 3^r + c_{r-1}3^{r-1} + \cdots + c_2 \cdot 3^2 + c_1 \cdot 3 + c_0$$

where r is a nonnegative integer, $c_r = 1$ or 2 and $c_j = 0$, 1, or 2 for all $j = 0, 1, 2, , \ldots, r - 1$.

We must show that k can be written in the required form. By the quotient-remainder theorem, k can be written as $3m$, $3m + 1$, or $3m + 2$ for some integer m. In each case m satisfies $0 \leq m < k$, and so m can be written in the required form:

$$m = c_r \cdot 3^r + c_{r-1}3^{r-1} + \cdots + c_2 \cdot 3^2 + c_1 \cdot 3 + c_0$$

where r is a nonnegative integer, $c_r = 1$ or 2 and $c_j = 0$, 1, or 2 for all $j = 0, 1, 2, , \ldots, r - 1$. Then

$$3m = c_r \cdot 3^{r+1} + c_{r-1}3^r + \cdots + c_2 \cdot 3^3 + c_1 \cdot 3^2 + c_0 \cdot 3 + 0$$
$$3m + 1 = c_r \cdot 3^{r+1} + c_{r-1}3^r + \cdots + c_2 \cdot 3^3 + c_1 \cdot 3^2 + c_0 \cdot 3 + 1$$
$$3m + 2 = c_r \cdot 3^{r+1} + c_{r-1}3^r + \cdots + c_2 \cdot 3^3 + c_1 \cdot 3^2 + c_0 \cdot 3 + 2$$

which all have the required form. Hence k can be written in the required form *[as was to be shown]*.

30. *Proof*: We first use the principle of ordinary mathematical induction to prove the well-ordering principle for the integers. Let S be a set of integers with one or more elements all of which are greater than or equal to some integer a, and suppose S does not have a least element. Let $P(n)$ be the property "$i \notin S$ for any integer i with $a \leq i \leq n$."

Show that the property is true for $n = a$: If a were in S, then it would be the least element of S because every element of S is greater than or equal to a. So, since S is assumed not to have a least element, $a \notin S$.

Show that for all integers $k \geq a$, if the property is true for $n = k$ then it is true for $n = k + 1$: Let an integer $k \geq a$ be given, and suppose that $i \notin S$ for any integer i with

$a \le i \le k$. *[This is the inductive hypothesis.]* It follows that if $k+1$ were an element of S, it would be the least element of S. But this is impossible because S is assumed not to have a least element. Thus $i \notin S$ for any integer i with $a \le i \le k+1$.

Hence, by ordinary mathematical induction, $i \notin S$ for any integer i with $a \le i \le n$, and, therefore, S does not contain any integer greater than or equal to a. But this contradicts the fact that S consists entirely of integers greater than or equal to a and has one or more elements. Thus the supposition that S does not have a least element is false, and so S has a least element.

This shows that the well-ordering principle for the integers follows from the principle of ordinary mathematical induction, and, therefore, that any statement that can be proved by the well-ordering principle for the integers can be proved by first using the principle of ordinary mathematical induction to deduce the well-ordering principle for the integers and then using the well-ordering principle for the integers to deduce the given statement.

Section 4.5

9. *Proof*:

I. Basis Property: $I(0)$ is the statement "both a and A are even integers or both are odd integers and $a \ge -1$." According to the pre-condition this statement is true.

II. Inductive Property: Suppose k is a nonnegative integer such that $G \wedge I(k)$ is true before an iteration of the loop. Then when execution comes to the top of the loop, $a_{old} > 0$ and a_{old} and A are both even integers or both are odd integers, and $a_{old} \ge -1$. Execution of statement 1 sets a_{new} equal to $a_{old} - 2$. Hence a_{new} has the same parity as a_{old} which is the same as A. Also since $a_{old} > 0$, then $a_{new} = a_{old} - 2 > 0 - 2 = -2$. But a_{new} is an integer. So since $a_{new} > -2$, $a_{new} \ge -1$. Hence after the loop iteration, $I(k+1)$ is true.

III. Eventual Falsity of Guard: The guard G is the condition $a > 0$. After each iteration of the loop, $a_{new} = a_{old} - 2 < a_{old}$, and so successive iterations of the loop give a strictly decreasing sequence of integer values of a which eventually becomes less than or equal to zero, at which point G becomes false.

IV. Correctness of the Post-Condition: Suppose that N is the least number of iterations after which G is false and $I(N)$ is true. Then (since G is false) $a \le 0$ and (since $I(N)$ is true) both a and A are even integers or both are odd integers, and $a \ge -1$. Putting the inequalities together gives $-1 \le a \le 0$, and so since a is an integer, $a = -1$ or $a = 0$. Since a and A have the same parity, then, $a = 0$ if A is even and $a = -1$ if A is odd. This is the post-condition.

12. *a*. Suppose the following condition is satisfied before entry to the loop: "there exist integers u, v, s, and t such that $a = uA + vB$ and $b = sA + tB$." Then

$$a_{old} = u_{old}A + v_{old}B \qquad \text{and} \qquad b_{old} = s_{old}A + t_{old}B,$$

for some integers u_{old}, v_{old}, s_{old}, and t_{old}. Observe that $b_{new} = r_{new} = a_{old} \bmod b_{old}$. So by the quotient-remainder theorem, there exists a unique integer q_{new} with $a_{old} = b_{old} \cdot q_{new} + r_{new} = b_{old} \cdot q_{new} + b_{new}$. Solving for b_{new} gives

$$\begin{aligned}
b_{new} &= a_{old} - b_{old} \cdot q_{new} \\
&= (u_{old}A + v_{old}B) - (s_{old}A + t_{old}B)q_{new} \\
&= (u_{old} - s_{old}q_{new})A + (v_{old} - t_{old}q_{new})B.
\end{aligned}$$

Therefore, let $s_{new} = u_{old} - s_{old}q_{new}$ and $t_{new} = v_{old} - t_{old}q_{new}$. Also since $a_{new} = b_{old} = s_{old}A + t_{old}B$, let $u_{new} = s_{old}$ and $v_{new} = t_{old}$. Hence $a_{new} = u_{new} \cdot A + v_{new} \cdot B$ and $b_{new} = s_{new} \cdot A + t_{new} \cdot B$, and so the condition is true after each iteration of the loop and hence after exit from the loop.

b. Initially $a = A$ and $b = B$. Let $u = 1$, $v = 0$, $s = 0$, and $t = 1$. Then before the first iteration of the loop, $a = uA + vB$ and $b = sA + tB$ as was to be shown.

c. By part (b) there exist integers u, v, s, and t such that before the first iteration of the loop, $a = uA + vB$ and $b = sA + tB$. So by part (a), after each subsequent iteration of the loop, there exist integers u, v, s, and t such that $a = uA + vB$ and $b = sA + tB$. Now after the final iteration of the **while** loop in the Euclidean algorithm, the variable *gcd* is given the current value of a (see page 196). But by the correctness proof for the Euclidean algorithm, $gcd = \gcd(A, B)$. Hence there exist integers u and v such that $\gcd(A, B) = uA + vB$.

d. The method discussed in part (a) gives the following formulas for u, v, s, and t:

$$u_{new} = s_{old}, \qquad v_{new} = t_{old}, \qquad s_{new} = u_{old} - s_{old}q_{new}, \qquad \text{and} \qquad t_{new} = v_{old} - t_{old}q_{new},$$

where in each iteration q_{new} is the quotient obtained by dividing a_{old} by b_{old}. The trace table below shows the values of a, b, r, q, gcd, and u, v, s, and t for the iterations of the **while** loop from the Euclidean algorithm. By part (b) the initial values of u, v, s, and t are $u = 1$, $v = 0$, $s = 0$, and $t = 1$.

r		18	12	6	0
q		2	8	1	2
a	330	156	18	12	6
b	156	18	12	6	0
gcd					6
u	1	0	1	-8	9
v	0	1	-2	17	-19
s	0	1	-8	9	-26
t	1	-2	17	-19	55

Since the final values of gcd, u, and v are 6, 9 and -19 and since $A = 330$ and $B = 156$, we have $\gcd(330, 156) = 6 = 330u + 156v = 330 \cdot 9 + 156 \cdot (-19)$, which is true.

General Review Guide: Chapter 4

Sequences and Summations

- What is a method to help find an explicit formula for a sequence whose first few terms are given (provided a nice explicit formula exists!)? *(p. 201)*
- What is the summation notation for a sum that is given in expanded form? *(p. 202)*
- What is the expanded form for a sum that is given in summation notation? *(p. 203)*
- What is the product notation? *(p. 205)*
- What is factorial notation? *(p. 206)*
- What are some properties of summations and products? *(p. 207)*
- How do you transform a summation by making a change of variable? *(p. 209)*
- What is an algorithm for converting from base 10 to base 2? *(p. 211)*

Mathematical Induction

- What do you show in the basis step and what do you show in the inductive step when you use (ordinary) mathematical induction to prove that a property involving an integer n is true for all integers greater than or equal to some initial integer? *(p. 218)*
- What is the inductive hypothesis in a proof by (ordinary) mathematical induction? *(p. 218)*
- Are you able to use (ordinary) mathematical induction to construct proofs involving various kinds of statements such as formulas, divisibility properties, and inequalities? *(pp. 218, 220, 223, 229, 231, 232)*
- Are you able to apply the formula for the sum of the first n positive integers? *(p. 222)*
- Are you able to apply the formula for the sum of the successive powers of a number, starting with the zeroth power? *(p. 225)*

Strong Mathematical Induction and The Well-Ordering Principle

- What do you show in the basis step and what do you show in the inductive step when you use strong mathematical induction to prove that a property involving an integer n is true for all integers greater than or equal to some initial integer? *(p. 235)*
- What is the inductive hypothesis in a proof by strong mathematical induction? *(p. 235)*
- Are you able to use strong mathematical induction to construct proofs of various statements? *(pp. 236-240)*
- What is the well-ordering principle for the integers? *(p. 240)*
- Are you able to use the well-ordering principle for the integers to prove statements, such as the existence part of the quotient-remainder theorem? *(p. 241)*
- How are ordinary mathematical induction, strong mathematical induction, and the well-ordering principle related? *(p. 240)*

Algorithm Correctness

- What are the pre-condition and the post-condition for an algorithm? *(p. 245)*
- What does it mean for a loop to be correct with respect to its pre- and post-conditions? *(p. 246)*
- What is a loop invariant? *(p. 247)*
- How do you use the loop invariant theorem to prove that a loop is correct with respect to its pre- and post-conditions? *(pp. 248-253)*

Test Your Understanding: Chapter 4

Test yourself by filling in the blanks.

1. The expanded form of the summation $\sum_{k=1}^{n} a_k$ is _____.

2. When $n = 1$, the value of $1^2 + 2^2 + 3^2 + \cdots + n^2$ is _____.

3. The expanded form of the product $\prod_{i=0}^{m} c_i$ is _____.

4. The notation $n! =$ _____.

5. When $c\sum_{k=1}^{n} a_k + \sum_{k=1}^{n} b_k$ is written as a single summation, the result is _____.

6. If you start with $\sum_{k=1}^{n} a_k$ and make the change of variable $j = k - 1$, the result is _____.

7. Repeated division by 2 is used to convert a positive integer to _____ notation.

8. To prove by (ordinary) mathematical induction that a property is true for all integers $n \geq a$, the first step is to show that _____ and the second step is to show that _____.

9. To prove by (ordinary) mathematical induction that $P(n)$ is true for all integers n greater than or equal to some integer a, the inductive hypothesis in the inductive step is _____.

10. To prove by strong mathematical induction that a property is true for all integers $n \geq a$, the first step is to show that _____ and the second step is to show that _____.

11. If the claim that $P(n)$ is true for all integers n greater than or equal to some integer a is proved by strong mathematical induction and if the basis step shows that $P(n)$ is true for all integers n with $a \leq n \leq b$, then the inductive hypothesis in the inductive step is _____.

12. The well-ordering principle for the integers says that _____.

13. A pre-condition for an algorithm is _____ and a post-condition for an algorithm is _____.

14. A loop is defined as correct with respect to its pre- and post-conditions if, and only if, whenever the algorithm variables satisfy the pre-condition for the loop and the loop terminates after a finite number of steps, then _____.

15. For each iteration of a loop, if a loop invariant is true before iteration of the loop, then _____.

16. Given a **while** loop with guard G and a predicate $I(n)$ if the following four properties are true, then the loop is correct with respect to its pre- and post-conditions:
    (1) The pre-condition for the loop implies that _____ is true before the first iteration of the loop;
    (2) For all integers $k \geq 0$, if the guard G and the predicate $I(k)$ are both true before an iteration of the loop, then _____;
    (3) After a finite number of iterations of the loop, _____;
    (4) If N is the least number of iterations after which G is false and $I(N)$ is true, then the values of the algorithm variables will be as specified _____.

Answers

1. $a_1 + a_2 + \cdots + a_n$

2. 1

3. $c_1 c_2 \cdots c_m$

4. $n(n-1)(n-2) \cdots 3 \cdot 2 \cdot 1$

5. $\displaystyle\sum_{k=1}^{n} (ca_k + b_k)$

6. $\displaystyle\sum_{j=0}^{n-1} a_{j+1}$

7. binary

8. the property is true for $n = a$

   for all integers $k \geq a$, if the property is true for $n = k$ then it is true for $n = k + 1$

9. the supposition that for any *[particular but arbitrarily chosen]* integer k with $k \geq a$, $P(k)$ is true

10. the property is true for an initial integer or set of initial integers

    for any integer k that is greater than the largest of the initial integers, if the property is true for all integers from the smallest of the initial integers through $k - 1$, then it is true for k

11. the supposition that for any *[particular but arbitrarily chosen]* integer with $k \geq b$, $P(i)$ is true for all integers i with $a \leq i < k$

12. any set of integers, all of which are greater than or equal to some fixed integer, has a least element

13. a predicate that describes the initial state of the input variables for the algorithm;

    a predicate that describes the final state of the output variables for the algorithm

14. the algorithm variables satisfy the post-condition for the loop

15. it is true after iteration of the loop

16. $I(0)$ is true

    $I(k + 1)$ is true after the iteration of the loop

    the guard G becomes false

    in the post-condition of the loop

Formats for Proving Formulas by Mathematical Induction

When using mathematical induction to prove a formula, students are sometimes tempted to present their proofs in a way that assumes what is to be proved. There are several formats you can use, besides the one shown most frequently in the textbook, to avoid this fallacy. A crucial point is this:

> If you are hoping to prove that an equation is true but you haven't yet done so, either preface it with the words "We must show that" or put a question mark above the equal sign.

Format 1 (the format used most often in the textbook for the inductive step): Start with the left-hand side (LHS) of the equation to be proved and successively transform it using definitions, known facts from basic algebra, and (for the inductive step) the inductive hypothesis until you obtain the right-hand side (RHS) of the equation.

Format 2 (the format used most often in the textbook for the basis step): Transform the LHS and the RHS of the equation to be proved *independently*, one after the other, until both sides are shown to equal the same expression. Because two quantities equal to the same quantity are equal to each other, you can conclude that the two sides of the equation are equal to each other.

Format 3: This format is just like Format 2 except that the computations are done in parallel. But in order to avoid the fallacy of assuming what is to be proved, do NOT put an equal sign between the two sides of the equation until the very last step. Separate the two sides of the equation with a vertical line.

Format 4: This format is just like Format 3 except that the two sides of the equation are separated by an equal sign with a question mark on top: $\stackrel{?}{=}$

Format 5: Start by writing something like "We must show that" and the equation you want to prove true. In successive steps, indicate that this equation is true if, and only if, ($\Leftrightarrow$) various other equations are true. But be sure that both the directions of your "if and only if" claims are correct. In other words, be sure that the $\Leftarrow$ direction is just as true as the $\Rightarrow$ direction. If you finally get down to an equation that is known to be true, then because each subsequent equation is true *if, and only if*, the previous equation is true, you will have shown that the original equation is true.

Example: Prove that for each integer $n \geq 1$,

$$\boxed{1 + 3 + 5 + \cdots + (2n - 1) = n^2} \leftarrow \text{This is the equation.}$$

Proof that the equation is true for $n = 1$:

Solution (Format 2):
When $n = 1$, the LHS of the equation equals 1, and the RHS equals 1^2 which also equals 1. So the equation is true for $n = 1$.

Solution (Format 5):
When $n = 2$, we must show that $1 = 1^2$. Because this is true, the equation is true for $n = 1$.

Proof that if the equation is true for $n = k$ then it is true for $n = k + 1$:

Solution (Format 2):

Suppose that for some integer $k \geq 1$, $1 + 3 + 5 + \cdots + (2k - 1) = k^2$. *[This is the inductive hypothesis.]*

We must show that $1 + 3 + 5 + \cdots + (2k + 1) = (k + 1)^2$.

But the LHS of the equation to be shown is

$$1 + 3 + 5 + \cdots + (2k + 1) = 1 + 3 + 5 + \cdots + (2k - 1) + (2k + 1)$$
$$\text{by making the next-to-last term explicit}$$
$$= k^2 + (2k + 1) \quad \text{by inductive hypothesis.}$$

And the RHS of the equation to be shown is

$$(k + 1)^2 = k^2 + 2k + 1 \qquad \text{by basic algebra.}$$

So the LHS and the RHS are equal to the same quantity, and thus they are equal to each other *[as was to be shown]*.

Solution (Format 3):

Suppose that for some integer $k \geq 1$, $1 + 3 + 5 + \cdots + (2k - 1) = k^2$. *[This is the inductive hypothesis.]*

We must show that $1 + 3 + 5 + \cdots + (2k + 1) = (k + 1)^2$.

But

$$
\begin{array}{l|l}
1 + 3 + 5 + \cdots + (2k + 1) & (k + 1)^2 \\
= 1 + 3 + 5 + \cdots + (2k - 1) + (2k + 1) & \\
\quad \text{by making the next-to-last term explicit} & \\
= k^2 + (2k + 1) & \\
\quad \text{by inductive hypothesis} & \\
= k^2 + 2k + 1 & = k^2 + 2k + 1 \\
\quad \text{by basic algebra} & \quad \text{by basic algebra}
\end{array}
$$

So the LHS and the RHS are equal to the same quantity, and thus they are equal to each other *[as was to be shown]*.

Solution (Format 4):

Suppose that for some integer $k \geq 1$, $1 + 3 + 5 + \cdots + (2k - 1) = k^2$. *[This is the inductive hypothesis.]*

We must show that $1 + 3 + 5 + \cdots + (2k + 1) = (k + 1)^2$.

But

$$
\begin{array}{ll}
1 + 3 + 5 + \cdots + (2k + 1) & \overset{?}{=} \quad (k + 1)^2 \\
1 + 3 + 5 + \cdots + (2k - 1) + (2k + 1) & \overset{?}{=} \quad k^2 + 2k + 1 \\
\quad \text{by making the next-to-last term explicit} & \qquad\qquad \text{by basic algebra} \\
k^2 + (2k + 1) & \overset{?}{=} \quad k^2 + 2k + 1 \\
\quad \text{by inductive hypothesis} & \\
k^2 + 2k + 1 & = \quad k^2 + 2k + 1 \\
\quad \text{by basic algebra} &
\end{array}
$$

So the LHS and the RHS are equal to the same quantity, and thus they are equal to each other *[as was to be shown]*.

Solution (Format 5):

Suppose that for some integer $k \geq 1$, $1 + 3 + 5 + \cdots + (2k - 1) = k^2$. *[This is the inductive hypothesis.]*

We must show that $1 + 3 + 5 + \cdots + (2k + 1) = (k + 1)^2$.

But this equation is true if, and only if, ($\Leftrightarrow$)

$$
\begin{array}{rrcll}
& 1 + 3 + 5 + \cdots + (2k - 1) + (2k + 1) & = & (k + 1)^2 & \text{by making the next-to-last term explicit} \\
\Leftrightarrow & k^2 + (2k + 1) & = & (k + 1)^2 & \text{by inductive hypothesis} \\
\Leftrightarrow & k^2 + 2k + 1 & = & (k + 1)^2 &
\end{array}
$$

which is true by basic algebra. Thus the equation to be shown is also true.

Chapter 5: Set Theory

The first section of this chapter introduces the structures of set and ordered set and illustrates them with a variety of examples. The aim of this section is to provide a solid basis of experience for deriving the set properties discussed in the remainder of the chapter. The second and third sections show how to prove and disprove various proposed set properties using element arguments, algebraic arguments, and counterexamples. Section 5.3 also shows that the concept of Boolean algebra generalizes both the algebra of sets with the operations of union and intersection and the properties of a set of statements with the operations of *or* and *and*. Section 5.4 discusses Russell's paradox and shows that reasoning similar to Russell's can be used to prove an important property of computer algorithms.

Section 5.1

3. $A = D$

6. *b. In words:* The set of all x in the universal set U such that x is in A or x is in B.

   Shorthand notation: $A \cup B$.

   c. In words: The set of all x in the universal set U such that x is in A and x is not in B.

   Shorthand notation: $A - B$.

   d. In words: The set of all x in the universal set U such that x is not in A.

   Shorthand notation: A^c.

9. *f.* $B - A = \{6\}$ *g.* $B \cup C = \{2, 3, 4, 6, 8, 9\}$ *h.* $B \cap C = \{6\}$

12. *a.* True: Every positive integer is a rational number.

   c. False: There are many rational numbers that are not integers. For instance, $1/2 \in \mathbf{Q}$ but $1/2 \notin \mathbf{Z}$.

   e. True: No integers are both positive and negative.

   f. True: Every rational number is real. So the set of all numbers that are both rational and real is the same as the set of all numbers that are rational.

   g. True: Every integer is a rational number, and so the set of all numbers that are integers or rational numbers is the same as the set of all rational numbers.

   h. True: Every positive integer is a real number, and so the set of all numbers that are both positive integers and real numbers is the same as the set of all positive integers.

   i. False: Every integer is a rational number, and so the set of all numbers that are integers or rational numbers is the same as the set of all rational numbers. However there are many rational numbers that are not integers, and so $\mathbf{Z} \cup \mathbf{Q} = \mathbf{Q} \neq \mathbf{Z}$.

15. *a.* No. For example, $5 \in A$ because $5 = 5 \cdot 1$, but $5 \neq 20k$ for any integer k, and so $5 \notin B$.

   b. Yes: If n is any element of B, then $n = 20k$ for some integer k. Thus $n = 5(4k)$ and so, since $4k$ is an integer, $n \in A$.

18. *b.*

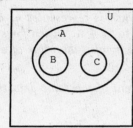 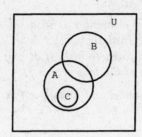 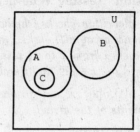

21.

   b   *c*   *d*

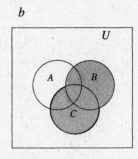 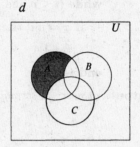

   e   *f*

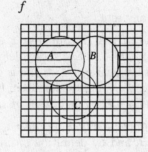

Region shaded $|\,|\,|$ *is* A^c
Region shaded $\equiv$ *is* B^c

Cross-hatched region is $A^c \cap B^c$

24. Yes. Every real number x satisfies exactly one of the conditions: $x > 0$ or $x = 0$ or $x < 0$. (See property T16 of Appendix A.)

27. *b.* $X \times Y = \{(a,x),(a,y),(b,x),(b,y)\}$

$\mathscr{P}(X \times Y) \ = \ \{\emptyset, \{(a,x)\}, \{(a,y)\}, \{(b,x)\}, \{(b,y)\}, \{(a,x),(a,y)\}, \{(a,x),(b,x)\},$
$\{(a,x),(b,y)\}, \{(a,y),(b,x)\}, \{(a,y),(b,y)\}, \{(b,x),(b,y)\},$
$\{(a,x),(a,y),(b,x)\}, \{(a,x),(a,y),(b,y)\}, \{(a,x),(b,x),(b,y)\},$
$\{(a,y),(b,x),(b,y)\}, \{(a,x),(a,y),(b,x),(b,y)\}\}$

28. *a.* $\mathscr{P}(\emptyset) = \{\emptyset\}$

   c. $\mathscr{P}(\mathscr{P}(\mathscr{P}(\emptyset))) = \{\emptyset, \{\emptyset\}, \{\{\emptyset\}\}, \{\emptyset, \{\emptyset\}\}\}$

30. *b.* $(A \times B) \times C = \{((1,u),m),((1,u),n),((1,v),m),((1,v),n),((2,u),m),((2,u),n),((2,v),m),$
$((2,v),n),((3,u),m),((3,u),n),((3,v),m),((3,v),n)\}$

   c. $A \times B \times C = \{(1,u,m),(1,u,n),(1,v,m),(1,v,n),(2,u,m),(2,u,n),(2,v,m),(2,v,n),$
$(3,u,m),(3,u,n),(3,v,m),(3,v,n)\}$

33. **Algorithm: Testing Whether $x \in A$**

    [This algorithm checks whether an element x is in a set A, which is represented as a one-dimensional array a[1],a[2],...,a[n]. Initially answer is set equal to "x ∉ A." Then for successive integers i from 1 to n, x is compared to a[i]. If at any stage x=a[i], the value of answer is changed to "x ∈ A" and iteration of the loop ceases.]

    Input: $a[1], a[2], \ldots, a[n]$ *[a one-dimensional array]*, x *[an element of the same data type as the elements of the array]*

    Algorithm Body:

    $\qquad i := 1, \; answer := \text{``}x \notin A\text{''}$

    $\qquad$ **while** $(i \leq n$ **and** $answer = \text{``}x \notin A\text{''})$

    $\qquad\quad$ **if** $x = a[i]$ **then** $answer := \text{``}x \in A\text{''}$

    $\qquad\quad i := i + 1$

    $\qquad$ **end while**

    Output: *answer* *[a string]*

Section 5.2

9. *Proof*: Let A, B, and C be any sets.

   $(A - B) \cap (C - B) \subseteq (A \cap C) - B$: Suppose $x \in (A - B) \cap (C - B)$. By definition of intersection, $x \in A - B$ and $x \in C - B$, and so, by definition of set difference, $x \in A$ and $x \notin B$ and $x \in C$ and $x \notin B$. To summarize: $x \in A$ and $x \in C$ and $x \notin B$. Hence, by definition of intersection, $x \in A \cap C$ and $x \notin B$, and by definition of set difference, $x \in (A \cap C) - B$. *[Thus $(A - B) \cap (C - B) \subseteq (A \cap C) - B$ by definition of subset.]*

   $(A \cap C) - B \subseteq (A - B) \cap (C - B)$: Suppose $x \in (A \cap C) - B$. By definition of set difference, $x \in (A \cap C)$ and $x \notin B$, and by definition of intersection, $x \in A$ and $x \in C$ and $x \notin B$. Thus it is true that $x \in A$ and $x \notin B$ and $x \in C$ and $x \notin B$, and so by definition of set difference, $x \in A - B$ and $x \in C - B$. Therefore by definition of intersection, $x \in (A - B) \cap (C - B)$. *[Thus $(A \cap C) - B \subseteq (A - B) \cap (C - B)$ by definition of subset.]*

   [Since both subset containments have been proved, $(A - B) \cap (C - B) = (A \cap C) - B$ by definition of set equality.]

15. *Proof*: Suppose A, B, and C are sets and $A \subseteq B$ and $A \subseteq C$. Let $x \in A$. Since $x \in A$ and $A \subseteq B$, then $x \in B$ (by definition of subset). Similarly, since $x \in A$ and $A \subseteq C$, then $x \in C$. Hence $x \in B$ and $x \in C$, and so by definition of intersection, $x \in B \cap C$. *[By definition of subset, therefore, $A \subseteq B \cap C$.]*

21.

darkly shaded region is A∩(B∪C)

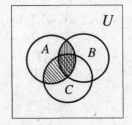

entire shaded region is (A∩B)∪(A∩C)

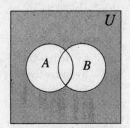

c

shaded region is $(A \cup B)^c$

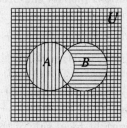

cross-hatched region is $A^c \cap B^c$

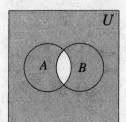

d

shaded region is $(A \cap B)^c$

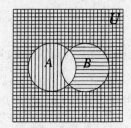

entire shaded region is $A^c \cup B^c$

24. *Proof*: Let A, B, and C be any sets, and suppose that $(A-B) \cap (B-C) \cap (A-C) \neq \emptyset$. Then there is an element x such that $x \in (A-B) \cap (B-C) \cap (A-C)$. By definition of intersection, $x \in A-B$ and $x \in B-C$ and $x \in A-C$, and so by definition of set difference, $x \in A$ and $x \notin B$ and $x \in B$ and $x \notin C$ and $x \in A$ and $x \notin C$. In particular, $x \notin B$ and $x \in B$, which is a contradiction. Hence the supposition is false. That is, $(A-B) \cap (B-C) \cap (A-C) = \emptyset$.

30. *Proof*: Let A, B, and C be any sets such that $A \subseteq B$ and $B \cap C = \emptyset$. Suppose $A \cap C \neq \emptyset$, that is, suppose there were an element x in $A \cap C$. Then by definition of intersection, $x \in A$ and $x \in C$. But by hypothesis $A \subseteq B$, and so since $x \in A$, $x \in B$ also. Hence $x \in B \cap C$, which implies that $B \cap C \neq \emptyset$. But this contradicts the hypothesis that $B \cap C = \emptyset$. Hence the supposition is false, and so $A \cap C = \emptyset$.

33. *Proof*: Let A, B, and C be any sets such that $B \cap C \subseteq A$. Suppose $(C-A) \cap (B-A) \neq \emptyset$. Then there is an element x such that $x \in (C-A) \cap (B-A)$. By definition of intersection, $x \in C-A$ and $x \in B-A$, and so by definition of set difference, $x \in C$ and $x \notin A$ and $x \in B$ and $x \notin A$. Since $x \in C$ and $x \in B$, $x \in B \cap C$ by definition of intersection. But $B \cap C \subseteq A$, and so $x \in A$. Thus $x \notin A$ and $x \in A$, which is a contradiction. Hence the supposition is false, and so $(C-A) \cap (B-A) = \emptyset$.

36. *Proof (by mathematical induction)*: Let the property $P(n)$ be the equation

$$(A_1 - B) \cup (A_2 - B) \cup \cdots \cup (A_n - B) = (A_1 \cup A_2 \cup \cdots \cup A_n) - B.$$

Show that the property is true for $n = 1$: For $n = 1$ the property is the equation $A_1 - B = A_1 - B$, which is clearly true.

Show that for all integers $k \geq 1$, if the property is true for $n = k$ then it is true for $n = k + 1$: Let k be an integer with $k \geq 1$, and suppose $A_1, A_2, \ldots, A_k, A_{k+1}$, and B are any sets such that

$$(A_1 - B) \cup (A_2 - B) \cup \cdots \cup (A_k - B) = (A_1 \cup A_2 \cup \cdots \cup A_k) - B. \qquad \leftarrow \text{inductive hypothesis}$$

We must show that

$$(A_1 - B) \cup (A_2 - B) \cup \cdots \cup (A_{k+1} - B) = (A_1 \cup A_2 \cup \cdots \cup A_{k+1}) - B.$$

But the left-hand side of this equation is

$$(A_1 - B) \cup (A_2 - B) \cup \cdots \cup (A_{k+1} - B)$$

$$
\begin{aligned}
&= && [(A_1 - B) \cup (A_2 - B) \cup \cdots \cup (A_k - B)] \cup (A_{k+1} - B) && \text{by definition} \\
&= && [(A_1 \cup A_2 \cup \cdots \cup A_k) - B] \cup (A_{k+1} - B) && \text{by inductive hypothesis} \\
&= && [(A_1 \cup A_2 \cup \cdots \cup A_k) \cup A_{k+1}] - B && \text{by exercise 8} \\
&= && (A_1 \cup A_2 \cup \cdots \cup A_{k+1}) - B && \text{by definition}
\end{aligned}
$$

and this is the right-hand side of the equation *[as was to be shown]*.

Section 5.3

12. False. *Counterexample*: Let $A = \{1\}$ and $B = \{2\}$. Then $A \cap B = \emptyset$ but $A \times B = \{(1,2)\} \neq \emptyset$.

15. True. *Proof*: Let A and B be sets and suppose $X \in \mathscr{P}(A) \cup \mathscr{P}(B)$. Then $X \in \mathscr{P}(A)$ or $X \in \mathscr{P}(B)$ *[by definition of union]*. In case $X \in \mathscr{P}(A)$, then $X \subseteq A$ *[by definition of power set]*, and so $X \subseteq A \cup B$ *[by definition of union]*. In case $X \in \mathscr{P}(B)$, then $X \subseteq B$ *[by definition of power set]*, and so $X \subseteq A \cup B$ *[by definition of union]*. Thus in either case, $X \subseteq A \cup B$, and so $X \in \mathscr{P}(A \cup B)$ *[by definition of power set]*. Hence $\mathscr{P}(A) \cup \mathscr{P}(B) \subseteq \mathscr{P}(A \cup B)$ *[by definition of subset]*.

18. *b. Negation*: $\forall$ sets S, $\exists$ a set T such that $S \cup T \neq \emptyset$. The negation is true. For example, given any set S, let $T = \{1\}$. Then $S \cup T = S \cup \{1\}$. Since $1 \in S \cup \{1\}$, $S \cup \{1\} \neq \emptyset$.

21. *d.* S_1 and S_2 each have eight elements.

    e. $S_1 \cup S_2$ has sixteen elements.

    f. $S_1 \cup S_2 = \mathscr{P}(A)$.

24. *e.* double complement law   *f.* distributive law   *g.* set difference law

27. *Proof*: Let sets A, B, and C be given. Then

$$
\begin{aligned}
(A - B) - C &= (A \cap B^c) \cap C^c && \text{by the set difference law (used twice)} \\
&= A \cap (B^c \cap C^c) && \text{by the associative law for } \cap \\
&= A \cap (B \cup C)^c && \text{by De Morgan's law} \\
&= A - (B \cup C) && \text{by the set difference law.}
\end{aligned}
$$

30. *Proof*: Let sets A and B be given. Then

$$
\begin{aligned}
(B^c \cup (B^c - A))^c &= (B^c \cup (B^c \cap A^c))^c && \text{by the set difference law} \\
&= (B^c)^c \cap (B^c \cap A^c)^c && \text{by De Morgan's law} \\
&= B \cap (B^c \cap A^c)^c && \text{by the double complement law} \\
&= B \cap ((B^c)^c \cup (A^c)^c) && \text{by De Morgan's law} \\
&= B \cap (B \cup A) && \text{by the double complement law (used twice)} \\
&= B && \text{by the absorption law.}
\end{aligned}
$$

33. *Proof*: Let A, B, and C be any sets. Then

$$(A - B) - (B - C)$$

$$
\begin{aligned}
&= (A \cap B^c) \cap (B \cap C^c)^c && \text{by the set difference law (used three times)} \\
&= (A \cap B^c) \cap (B^c \cup (C^c)^c) && \text{by De Morgan's law} \\
&= (A \cap B^c) \cap (B^c \cup C) && \text{by the double complement law} \\
&= ((A \cap B^c) \cap B^c) \cup ((A \cap B^c) \cap C) && \text{by the distributive law} \\
&= (A \cap (B^c \cap B^c)) \cup ((A \cap B^c) \cap C) && \text{by the associative law for } \cap \\
&= (A \cap B^c) \cup ((A \cap B^c) \cap C) && \text{by the idempotent law for } \cap \\
&= A \cap B^c && \text{by the absorption law} \\
&= A - B && \text{by the set difference law.}
\end{aligned}
$$

36. Let A, B, and C be any sets. Then $((A \cap (B \cup C)) \cap (A - B)) \cap (B \cup C^c)$

$$
\begin{aligned}
&= ((A \cap (B \cup C)) \cap (A \cap B^c)) \cap (B \cup C^c) &&\text{by the set difference law} \\
&= ((A \cap B^c) \cap (A \cap (B \cup C))) \cap (B \cup C^c) &&\text{by the commutative law for } \cap \\
&= (((A \cap B^c) \cap A) \cap (B \cup C)) \cap (B \cup C^c) &&\text{by the associative law for } \cap \\
&= ((A \cap (A \cap B^c)) \cap (B \cup C)) \cap (B \cup C^c) &&\text{by the commutative law for } \cap \\
&= (((A \cap A) \cap B^c) \cap (B \cup C)) \cap (B \cup C^c) &&\text{by the associative law for } \cap \\
&= ((A \cap B^c) \cap (B \cup C)) \cap (B \cup C^c) &&\text{by the idempotent law for } \cap \\
&= (A \cap B^c) \cap ((B \cup C) \cap (B \cup C^c)) &&\text{by the associative law for } \cap \\
&= (A \cap B^c) \cap (B \cup (C \cap C^c)) &&\text{by the distributive law} \\
&= (A \cap B^c) \cap (B \cup \emptyset) &&\text{by the complement law for } \cap \\
&= (A \cap B^c) \cap B &&\text{by the identity law for } \cup \\
&= A \cap (B^c \cap B) &&\text{by the associative law for } \cap \\
&= A \cap (B \cap B^c) &&\text{by the commutative law for } \cap \\
&= A \cap \emptyset &&\text{by the complement law for } \cap \\
&= \emptyset &&\text{by the universal bound law for } \cap.
\end{aligned}
$$

39. *b.* $B \triangle C = (\{3,4,5,6\} - \{5,6,7,8\}) \cup (\{5,6,7,8\} - \{3,4,5,6\}) = \{3,4\} \cup \{7,8\} = \{3,4,7,8\}$

c. $A \triangle C = (\{1,2,3,4\} - \{5,6,7,8\}) \cup (\{5,6,7,8\} - \{1,2,3,4\}) = \{1,2,3,4\} \cup \{5,6,7,8\} = \{1,2,3,4,5,6,7,8\}$

d. By part (a), $A \triangle B = \{1,2,5,6\}$. So $(A \triangle B) \triangle C = (\{1,2,5,6\} - \{5,6,7,8\}) \cup (\{5,6,7,8\} - \{1,2,5,6\}) = \{1,2,7,8\} \cup \{7,8,1,2\} = \{1,2,7,8\}$.

42. *Proof*: Let A be any subset of a universal set U. Then

$$
\begin{aligned}
A \triangle A^c &= (A - A^c) \cup (A^c - A) &&\text{by definition of } \triangle \\
&= (A \cap (A^c)^c) \cup (A^c \cap A^c) &&\text{by the set difference law} \\
&= (A \cap A) \cup (A^c \cap A^c) &&\text{by the double complement law} \\
&= A \cup A^c &&\text{by the idempotent law for } \cap \text{ (used twice)} \\
&= U &&\text{by the complement law for } \cup.
\end{aligned}
$$

45. *Proof 1*: Suppose A, B, and C are any subsets of a universal set U. Then

$x \in (A \triangle B) \triangle C \quad \Leftrightarrow \quad (x \in A \triangle B \text{ and } x \notin C) \text{ or } (x \in C \text{ and } x \notin A \triangle B)$
$$\text{by the lemma from the solution to 44}$$

$\Leftrightarrow ([(x \in A \text{ and } x \notin B) \text{ or } (x \in B \text{ and } x \notin A)] \text{ and } x \notin C)$
$\text{or } (x \in C \text{ and } [(x \in A \text{ and } x \in B) \text{ or } (x \notin A \text{ and } x \notin B)])$
$$\text{by the lemma from the solution to 44}$$

$\Leftrightarrow ([x \in A \text{ and } x \notin B \text{ and } x \notin C] \text{ or } [x \in B \text{ and } x \notin A \text{ and } x \notin C])$
$\text{or } ([x \in C \text{ and } x \in A \text{ and } x \in B] \text{ or } [x \in C \text{ and } x \notin A \text{ and } x \notin B])$
$$\text{by the distributive and associative laws of logic}$$

$\Leftrightarrow x$ is in exactly one of the sets A, B, and C,
or x is in all three of the sets A, B, and C.

On the other hand,

$x \in A \triangle (B \triangle C) \quad \Leftrightarrow \quad (x \in A \text{ and } x \notin B \triangle C) \text{ or } (x \in B \triangle C \text{ and } x \notin A)$
$$\text{by the lemma from the solution to 44}$$

$\Leftrightarrow (x \in B \triangle C \text{ and } x \notin A) \text{ or } (x \in A \text{ and } x \notin B \triangle C)$
$$\text{by the commutative law for } or.$$

By exactly the same sequence of steps as in the first part of this proof but with B in place of A, C in place of B, and A in place of C, we deduce that

$x \in A \triangle (B \triangle C) \quad \Leftrightarrow \quad x$ is in exactly one of the sets A, B, and C, or
x is in all three of the sets A, B, and C.

So $x \in (A \triangle B) \triangle C \Leftrightarrow x \in A \triangle (B \triangle C)$, and hence $(A \triangle B) \triangle C = A \triangle (B \triangle C)$.

Proof 2: Suppose A, B, and C are any subsets of a universal set U. Then

$(A \triangle B) \triangle C$

$= ((A \triangle B) - C) \cup (C - (A \triangle B))$
> by definition of symmetric difference

$= (((A - B) \cup (B - A)) - C) \cup (C - ((A - B) \cup (B - A)))$
> by definition of symmetric difference (used twice)

$= (((A \cap B^c) \cup (B \cap A^c)) \cap C^c) \cup (C \cap ((A \cap B^c) \cup (B \cap A^c))^c)$
> by the set difference law (used six times)

$= (((A \cap B^c) \cap C^c) \cup ((B \cap A^c) \cap C^c))) \cup (C \cap ((A \cap B^c) \cup (B \cap A^c))^c)$
> by the commutative law for $\cap$ and the distributive law

$= (((A \cap B^c) \cap C^c) \cup ((A^c \cap B) \cap C^c)) \cup (C \cap ((A \cap B^c)^c \cap (B \cap A^c)^c))$
> by the commutative law for $\cap$ and De Morgan's law

$= (((A \cap B^c) \cap C^c) \cup ((A^c \cap B) \cap C^c)) \cup (C \cap ((A^c \cup B) \cap (B^c \cup A)))$
> by De Morgan's law and the double complement law

$= ((A \cap B^c \cap C^c) \cup (A^c \cap B \cap C^c)) \cup ((C \cap (A^c \cup B)) \cap (B^c \cup A))$
> by the associative law for $\cap$

$= ((A \cap B^c \cap C^c) \cup (A^c \cap B \cap C^c)) \cup (((C \cap A^c) \cup (C \cap B)) \cap (B^c \cup A))$
> by the distributive law

$= ((A \cap B^c \cap C^c) \cup (A^c \cap B \cap C^c)) \cup (((C \cap A^c) \cap (B^c \cup A)) \cup ((C \cap B) \cap (B^c \cup A)))$
> by the commutative law for $\cap$ and the distributive law

$= ((A \cap B^c \cap C^c) \cup (A^c \cap B \cap C^c)) \cup$
$\quad (((C \cap A^c) \cap B^c) \cup ((C \cap A^c) \cap A)) \cup ((C \cap B) \cap B^c) \cup ((C \cap B) \cap A)))$
> by the distributive law (used twice)

$= ((A \cap B^c \cap C^c) \cup (A^c \cap B \cap C^c)) \cup$
$\quad (((A^c \cap B^c \cap C) \cup (C \cap (A^c \cap A))) \cup ((C \cap (B \cap B^c)) \cup (A \cap B \cap C)))$
> by the commutative and associative laws for $\cap$

$= ((A \cap B^c \cap C^c) \cup (A^c \cap B \cap C^c)) \cup$
$\quad (((A^c \cap B^c \cap C) \cup (C \cap \emptyset)) \cup ((C \cap \emptyset) \cup (A \cap B \cap C)))$
> by the complement law for $\cap$ (used twice)

$= ((A \cap B^c \cap C^c) \cup (A^c \cap B \cap C^c)) \cup (((A^c \cap B^c \cap C) \cup \emptyset) \cup (\emptyset \cup (A \cap B \cap C)))$
> by the universal bound law for $\cap$ (used twice)

$= ((A \cap B^c \cap C^c) \cup (A^c \cap B \cap C^c)) \cup ((A^c \cap B^c \cap C) \cup (A \cap B \cap C))$
> by the commutative and identity laws for $\cup$

$= (A \cap B^c \cap C^c) \cup (A^c \cap B \cap C^c) \cup (A^c \cap B^c \cap C) \cup (A \cap B \cap C)$
> by the associative law for $\cup$.

A similar set of steps shows that $A \triangle (B \triangle C) = (A \cap B^c \cap C^c) \cup (A^c \cap B \cap C^c) \cup (A^c \cap B^c \cap C) \cup (A \cap B \cap C)$ also. Hence $(A \triangle B) \triangle C = A \triangle (B \triangle C)$.

54. *Proof*: By the uniqueness of the complement law, to show that $\bar{1} = 0$, it suffices to show that $1 + 0 = 1$ and $1 \cdot 0 = 0$. But the first equation is true by the identity law for $+$, and the second equation is true by exercise 51 (the universal bound law for $\cdot$). Thus $\bar{1} = 0$.

57. *Proof*: Let x, y, and z be any elements in B such that $x + y = x + z$ and $x \cdot y = x \cdot z$. Then

$$
\begin{aligned}
y &= (y + x) \cdot y && \text{by exercise 50} \\
&= (x + y) \cdot y && \text{by the commutative law for } + \\
&= (x + z) \cdot y && \text{by hypothesis} \\
&= y \cdot (x + z) && \text{by the commutative law for } \cdot \\
&= y \cdot x + y \cdot z && \text{by the distributive law for } \cdot \text{ over } + \\
&= (x \cdot y) + (z \cdot y) && \text{by the commutative law for } \cdot \text{ (used twice)} \\
&= (x \cdot z) + (z \cdot y) && \text{by hypothesis} \\
&= (z \cdot x) + (z \cdot y) && \text{by the commutative law for } \cdot \\
&= z \cdot (x + y) && \text{by the distributive law for } \cdot \text{ over } + \\
&= z \cdot (x + z) && \text{by hypothesis} \\
&= (z + x) \cdot z && \text{by the commutative laws for } \cdot \text{ and } + \\
&= z && \text{by exercise 50.}
\end{aligned}
$$

Section 5.4

3. This statement contradicts itself. If it were true, then because it declares itself to be a lie, it would be false. In other words, it is not true. On the other hand, if it were false, then it would be false that "the sentence in this box is a lie," and so the sentence would be true. In other words, the sentence is not false. Thus the sentence is neither true nor false, which contradicts the definition of a statement. Hence the sentence is not a statement.

6. In order for an *and* statement to be true, both components must be true. So if the given sentence is a true statement, the first component "this sentence is false" is true. But this implies that the sentence is false. Consequently, the sentence is not true. On the other hand, if the sentence is false, then at least one component is false. But because the second component "$1 + 1 = 2$" is true, the first component must be false. Thus it is false that "this sentence is false," and so the sentence is true. Consequently, the sentence is not false. Thus the sentence is neither true nor false, which contradicts the definition of a statement. Hence the sentence is not a statement.

12. Because the total number of strings consisting of 11 or fewer English words is finite, the number of such strings that describe integers must be also finite. Thus the number of integers described by such strings must be finite, and hence there is a largest such integer, say m. Let $n = m + 1$. Then n is "the smallest integer not describable in fewer than 12 English words." But this description of n contains only 11 words. So n is describable in fewer than 12 English words, which is a contradiction. (*Comment*: This contradiction results from the self-reference in the description of n.)

General Review for Chapter 5

Definitions: Can you define the following terms, use them correctly in sentences, and work with concrete examples involving them?

- subset *(p. 256)*
- proper subset *(p. 257)*
- equality of sets *(p. 258)*
- union, intersection, and difference of sets *(p. 260)*
- complement of a set *(p. 260)*
- empty set *(p. 262)*
- disjoint sets *(p. 262)*
- mutually disjoint sets *(p. 262)*
- partition of a set *(p. 263)*
- power set of a set *(p. 264)*
- Cartesian product of sets *(p. 265)*

Set Theory

- What is the difference between $\in$ and $\subseteq$? *(p. 258)*
- How do you use an element argument to prove that one set is a subset of another set? *(p. 269)*
- How are the procedural versions of set definitions used to prove properties of sets? *(p. 270)*
- What is the basic (two-step) method for showing that two sets are equal? *(p. 273)*
- Are you familiar with the set properties in Theorems 5.2.1 and 5.2.2? *(pp. 269, 272)*
- Why is the empty set a subset of every set? *(p. 278)*
- What is the special method used to show that a set equals the empty set? *(p. 279)*
- How do you find a counterexample for a proposed set identity? *(p. 283)*
- How do you find the number of subsets of a set with a finite number of elements? *(p. 285)*
- What is an "algebraic method" for proving that one set equals another set? *(p. 286)*
- What is a Boolean algebra? *(p. 288)*
- How do you deduce additional properties of a Boolean algebra from the properties that define it? *(p. 289)*
- What is Russell's paradox? *(p. 293)*
- What is the Halting Problem? *(p. 295)*

Test Your Understanding: Chapter 5

Test yourself by filling in the blanks.

1. The notation $x \in A$ is read ____.

2. The notation $A \subseteq B$ is read ____ and means that ____.

3. A set A equals a set B if, and only if, A and B have ____.

4. An element x is in $A \cup B$ if, and only if, ____.

5. An element x is in $A \cap B$ if, and only if, ____.

6. An element x is in $A - B$ if, and only if, ____.

7. An element x is in A^c if, and only if, ____.

8. The empty set is a set with ____.

9. The power set of a set A is ____.

10. Sets A and B are disjoint if, and only if, ____.

11. A collection of nonempty sets $A_1, A_2, \ldots, A_n$ is a partition of a set A if, and only if, ____.

12. Given sets A and B, the Cartesian product of A and B, $A \times B$, is ____.

13. Given sets $A_1, A_2, \ldots, A_n$, the Cartesian product $A_1 \times A_2 \times \cdots \times A_n$ is ____.

14. To use an element argument for proving that a set X is a subset of a set Y, you suppose that ____ and show that ____.

15. To use the basic method for proving that two sets X and Y are equal, you prove that ____ and that ____.

16. To prove a proposed set identity involving set variables A, B, and C, you suppose that ____ and show that ____.

17. If $\emptyset$ is a set with no elements and A is any set, the relation of $\emptyset$ and A is that ____.

18. To use the element method for proving that a set X equals the empty set, you prove that X has ____. To do this, you suppose that ____ and you show that this supposition leads to ____.

19. To show that a set X is not a subset of a set Y, ____.

20. Given a proposed set identity involving set variables A, B, and C, the most common way to show that the proposed identity is false is to find ____.

21. When using the "algebraic" method for proving a set identity, it is important to ____.

22. The operations of $+$ and $\cdot$ in a Boolean algebra are generalizations of the operations of ____ and ____ in the set of all statements forms in a given finite number of variables and the operations of ____ and ____ in the set of all subsets of a given set.

23. Russell showed that the following proposed "set definition" could not actually define a set: ____.

24. Turing's solution to the halting problem showed that there is no computer algorithm that will accept any algorithm X and data set D as input and then will indicate whether or not ____.

Answers

1. x is an element of the set A

2. the set A is a subset of the set B;
   for all x, if $x \in A$ then $x \in B$ (in other words, every element of A is also an element of B)

3. exactly the same elements

4. x is in A or x is in B

5. x is in A and x is in B

6. x is in A and x is not in B

7. x is not in A

8. no elements

9. the set of all subsets of A

10. $A \cap B = \emptyset$ (in other words, A and B have no elements in common)

11. $A = A_1 \cup A_2 \cup \cdots \cup A_n$ and $A_i \cap A_j = \emptyset$ for all $i, j = 1, 2, \ldots, n$ (in other words, A is the union of all the sets $A_1, A_2, \ldots, A_n$ and no two of these sets have any elements in common)

12. the set of all ordered pairs (a, b), where a is in A and b is in B

13. the set of all ordered n-tuples $(a_1, a_2, \ldots, a_n)$, where a_i is in A_i for all $i = 1, 2, \ldots, n$

14. x is any *[particular but arbitrarily chosen]* element of X
    x is an element of Y

15. $X \subseteq Y; Y \subseteq X$

16. $A, B,$ and C are any *[particular but arbitrarily chosen]* sets; the left-hand and right-hand sides of the equation are equal for those sets

17. $\emptyset \subseteq A$

18. no elements; there is at least one element in X; a contradiction

19. show that there is an element of X that is not an element of Y

20. concrete sets $A, B,$ and C for which the left-hand and right-hand sides of the equation are not equal

21. use the set properties from Theorem 5.2.2 exactly as they are stated

22. $\vee; \wedge; \cup; \cap$

23. the set of all sets that are not elements of themselves

24. execution of the algorithm terminates in a finite number of steps

Chapter 6: Counting

The primary aim of this chapter is to foster intuitive understanding for fundamental principles of counting and probability and an ability to apply them in a wide variety of situations. It is helpful to get into the habit of beginning a counting problem by listing (or at least imagining) some of the objects you are trying to count. If you see that all the objects to be counted can be matched up with the integers from m to n inclusive, then the total is $n - m + 1$ (Section 6.1). If you see that all the objects can be produced by a multi-step process, then the total can be found by counting the distinct paths from root to leaves in a possibility tree that shows the outcomes of each successive step (Section 6.2). And if each step of the process can be performed in a fixed number of ways (regardless of how the previous steps were performed), then the total can be calculated by applying the multiplication rule (Section 6.2).

If the objects to be counted can be separated into disjoint categories, then the total is just the sum of the subtotals for each category (Section 6.3). And if the categories are not disjoint, the total can be counted using the inclusion/exclusion rule (Section 6.3). If the objects to be counted can be represented as all the subsets of size r of a set with n elements, then the total is $\binom{n}{r}$ for which there is a computational formula (Section 6.4). Finally if the objects can be represented as all the multisets of size r of a set with n elements, then the total is $\binom{n+r-1}{r}$.

Pascal's formula is discussed in Section 6.6 and the binomial theorem in Section 6.7. Each is proved both algebraically and combinatorially. Exercise 21 of Section 6.4 is a warm-up for the combinatorial proof of the binomial theorem, and exercise 10 of Section 6.7 is intended to help you see how Pascal's formula is applied in the algebraic proof of the binomial theorem.

Sections 6.8 and 6.9 develop the axiomatic theory of probability through the concepts of expected value, conditional probability, independence, and Bayes theorem.

Section 6.1

6. $\{2\clubsuit, 3\clubsuit, 4\clubsuit, 2\diamondsuit, 3\diamondsuit, 4\diamondsuit, 2\heartsuit, 3\heartsuit, 4\heartsuit, 2\spadesuit, 3\spadesuit, 4\spadesuit\}$ Probability $= 12/52 = 3/13 \cong 23.1\%$

12. b. (ii) $\{GGB, GBG, BGG, GGG\}$ Probability $= 4/8 = 1/2 = 50\%$

    (iii) $\{BBB\}$ Probability $= 1/8 = 12.5\%$

15. The methods used to compute the probabilities in exercises 12, 13, and 14 are exactly the same as those in exercise 11. The only difference in the solutions are the symbols used to denote the outcomes; the probabilities are identical. These exercises illustrate the fact that computing various probabilities that arise in connection with tossing a coin is mathematically identical to computing probabilities in other, more realistic situations. So if the coin tossing model is completely understood, many other probabilities can be computed without difficulty.

27. Let k be the 62nd element in the array. By Theorem 6.1.1, $k - 29 + 1 = 62$, so $k = 62 + 29 - 1 = 90$. Thus the 62nd element in the array is $B[90]$.

30.

1	2	3	4	5	6	...	998	999	1000	1001
	$\updownarrow$		$\updownarrow$		$\updownarrow$		$\updownarrow$		$\updownarrow$	
	$2 \cdot 1$		$2 \cdot 2$		$2 \cdot 3$		$2 \cdot 499$		$2 \cdot 500$	

The diagram above shows that there are as many even integers between 1 and 1001 as there are integers from 1 to 500 inclusive. There are 500 such integers.

33. *Proof*: Let m be any integer, and let the property $P(n)$ be the sentence "The number of integers from m to n inclusive is $n - m + 1$." We will prove by mathematical induction that the property is true for all integers $n \geq m$.

    Show that the property is true for $n = m$: There is just one integer, namely m, from m to m inclusive. Substituting m in place of n in the formula $n - m + 1$ gives $m - m + 1 = 1$, which is correct.

    Show that for all integers $k \geq m$, if the property is true for $n = k$ then it is true for $n = k + 1$: Suppose k is an integer with $k \geq m$, and suppose the number of integers from m to k inclusive is $k - m + 1$. *[This is the inductive hypothesis.]* We must show that the number of integers from m to $k + 1$ inclusive is $m - (k + 1) + 1$.

    Consider the sequence of integers from m to $k + 1$ inclusive:

    $$\underbrace{m, \quad m + 1, \quad m + 2, \quad \ldots, \quad k,}_{k - m + 1} \quad (k + 1).$$

    By inductive hypothesis there are $k - m + 1$ integers from m to k inclusive. So there are $(k - m + 1) + 1$ integers from m to $k + 1$ inclusive. But $(k - m + 1) + 1 = (k + 1) - m + 1$. So there are $(k + 1) - m + 1$ integers from m to $k + 1$ inclusive *[as was to be shown]*.

Section 6.2

12. *b.* Think of creating a string of hexadecimal digits that satisfies the given requirements as a 6-step process. Step 1 is to choose the first hexadecimal digit. It can be any hexadecimal digit from 4 through D (which equals 13) There are $13 - 4 + 1 = 10$ of these. Steps 2–5 are to choose the second through the fifth hexadecimal digits. Each can be any one of the 16 hexadecimal digits. Step 6 is to choose the first hexadecimal digit. It can be any hexadecimal digit from 2 through (which equals 14) There are $14 - 2 + 1 = 11$ of these. So the total number of the specified hexadecimal numbers is $10 \cdot 16 \cdot 16 \cdot 16 \cdot 11 = 450,560$.

15. Think of creating combinations that satisfy the given requirements as multi-step processes in which steps 1-3 are to choose a number from 1 to 30, inclusive.

    a. Because there are 30 choices of numbers in each of steps 1–3, there are $30^3 = 27,000$ possible combinations for the lock.

    b. In this case we are given that no number may be repeated. So there are 30 choices for step 1, 29 for step 2, and 28 for step 3. Thus there are $30 \cdot 29 \cdot 28 = 24,360$ possible combinations for the lock.

18. *b.* There are 10 ways to perform step one, 22 ways to perform step two *[because we may choose any of the thirteen letters from N through Z or any of the nine digits not chosen in step 1]*, 34 ways to perform step three *[because we may not use either of the two previously used symbols]*, and 33 ways to perform step four (because we may not use any of the three previously used symbols). So the total number of PINs is $10 \cdot 22 \cdot 34 \cdot 33 = 246,840$.

27. *a.* Call one of the integers r and the other s. Since r and s have no common factors, if p_i is a factor of r, then p_i is not a factor of s. So for each $i = 1, 2, \ldots, m$, either $p_i^{k_i}$ is a factor of r or $p_i^{k_i}$ is a factor of s. Thus, constructing r can be thought of as an m-step process in which step i is to decide whether $p_i^{k_i}$ is a factor of r or not. There are two ways to perform each step, and so the number of different possible r's is 2^m. Observe that once r is specified, s is completely determined because $s = n/r$. Hence the number of ways n can be written as a product of two positive integers rs which have no common factors is 2^m. Note that this analysis assumes that order matters because, for instance, $r = 1$ and $s = n$ will be counted separately from $r = n$ and $s = 1$.

b. Each time that we can write n as rs, where r and s have no common factors, we can also write $n = sr$. So if order matters, there are twice as many ways to write n as a product of two integers with no common factors as there are if order does not matter. Thus if order does not matter, there are $2^m/2 = 2^{m-1}$ ways to write n as a product of two integers with no common factors.

30. *a.* The number of ways the 6 people can be seated equals the number or permutations of a set of 6 elements, namely, $6! = 720$.

b. Assuming that the row is bounded by two aisles, the answer is $2 \cdot 5! = 240$. Under this assumption, arranging the people in the row can be regarded as a 2-step process where step 1 is to choose the aisle seat for the doctor *[there are 2 ways to do this]* and step 2 is to choose an ordering for the remaining people *[there are 5! ways to do this]*. (If it is assumed that one end of the row is against a wall, then there is only one aisle seat and the answer is $5! = 120$.)

c. Each married couple can be regarded as a single item, so the number of ways to order the 3 couples is $3! = 6$.

33. $stu, stv, sut, suv, svt, svu, tsu, tsv, tus, tuv, tvs, tvu, ust, usv, uts, utv, uvs, uvt, vst, vsu, vts, vtu, vus, vut$

36. *b.* $P(9,6) = 9!/(9-6)! = 9!/3! = 9 \cdot 8 \cdot 7 \cdot 6 \cdot 5 \cdot 4 = 60,480$

d. $P(7,4) = 7!/(7-4)! = 7!/3! = 7 \cdot 6 \cdot 5 \cdot 4 = 840$

39. *Proof 1*: Let n be any integer such that $n \geq 3$. By Theorem 6.2.3,

$$
\begin{aligned}
P(n+1, 3) - P(n, 3) &= \frac{(n+1)!}{((n+1)-3)!} - \frac{n!}{(n-3)!} \\
&= \frac{(n+1)!}{(n-2)!} - \frac{n!}{(n-3)!} \\
&= \frac{(n+1) \cdot n!}{(n-2)!} - \frac{(n-2) \cdot n!}{(n-2) \cdot (n-3)!} \\
&= \frac{n!((n+1)-(n-2))}{(n-2)!} \\
&= \frac{n!}{(n-2)!} \cdot 3 \\
&= 3P(n, 2).
\end{aligned}
$$

Proof 2: Let n be any integer such that $n \geq 3$. By Theorem 6.2.3,

$$
\begin{aligned}
P(n+1, 3) - P(n, 3) &= (n+1)n(n-1) - n(n-1)(n-2) \\
&= n(n-1)[(n+1)-(n-2)] \\
&= n(n-1)(n+1-n+2) \\
&= 3n(n-1) \\
&= 3P(n, 2).
\end{aligned}
$$

42. *Proof*: For each integer $n \geq 1$, let the property $P(n)$ be the sentence "The number of permutations of a set with n elements is $n!$." We will prove by mathematical induction that the property is true for all integers $n \geq 1$.

Show that the property is true for $n = 1$: If a set consists of one element there is just one way to order it, and $1! = 1$.

Show that for all integers $k \geq 1$, if the property is true for $n = k$ then it is true for $n = k + 1$: Let k be an integer with $k \geq 1$ and suppose that the number of permutations of a set with k elements is $k!$. *[This is the inductive hypothesis.]* Let X be a set with $k + 1$ elements. The process of forming a permutation of the elements of X can be considered a two-step operation where step 1 is to choose the element to write first and step 2 is to write the remaining elements of X in some order. Since X has $k + 1$ elements, there are $k + 1$ ways to perform step 1, and by inductive hypothesis there are $k!$ ways to perform step 2. Hence by the product rule there are $(k + 1)k! = (k + 1)!$ ways to form a permutation of the elements of X. But this means that there are $(k + 1)!$ permutations of X *[as was to be shown]*.

Section 6.3

15. *a.* Imagine the process of constructing a string of four distinct hexadecimal digits as a 4-step operation:

    Step 1: Choose a hexadecimal digit to place into the first position in the string.

    Step 2: Choose another hexadecimal digit to place into the second position in the string.

    Step 3: Choose yet another hexadecimal digit to place into the third position in the string.

    Step 4: Choose yet another hexadecimal digit to place into the fourth position in the string.

    The total number of ways to construct the string (and hence the total number of such strings) is $16 \cdot 15 \cdot 14 \cdot 13 = 43,680$.

    b. The process of constructing an arbitrary string of four hexadecimal digits (including those in which not all the digits are distinct) can also be regarded as a 4-step operation with 16 ways to perform each step. So the total number of strings of four hexadecimal digits is $16^4 = 65,536$. Note that if not all the digits in a string are distinct, then at least one is repeated, and so by the difference rule, we may subtract the number that consist of four distinct digits from the total to obtain the number with at least one repeated digit. Thus the answer is $16^4 - 16 \cdot 15 \cdot 14 \cdot 13 = 21,856$.

    c. Because there are $16^4 = 65,536$ strings of four hexadecimal digits and because, by part (b), 21,856 of these have at least one repeated digit, the probability that a randomly chosen string of four hexadecimal digits has at least one repeated digit is $21856/65536 \cong 33.3\%$.

18. *b.* As in part (a), represent each integer from 1 through 99,999 as a string of five digits. The number of integers from 1 through 99,999 that do not contain the digit 6 is $9^5 - 1$ because there are 9 choices of digit for each of the five positions (namely, all ten digits except 6), except that 000000 is excluded. In addition, 100,000 does not contain the digit 6. So there are $(9^5 - 1) + 1 = 9^5$ integers from 1 through 100,000 that do not contain the digit 6. Therefore, by the difference rule, there are $100,000 - 9^5 = 40,951$ integers from 1 through 100,000 that contain at least one occurrence of the digit 6.

    c. By parts (a) and (b) and the difference rule, the number of integers from 1 through 100,000 that contain two or more occurrences of the digit 6 is the difference between the number that contain at least one occurrence and the number that contain exactly one occurrence, namely, $40,951 - 32,805 = 8146$. Since there are 100,000 integers from 1 through 100,000, the probability that a randomly chosen integer in this range contains two or more occurrences of the digit 6 is $8146/100000 = 8.146\%$.

21. *b.* By part (a) and the equally likely probability formula, the probability that an integer chosen at random from 1 through 1000 is a multiple of 4 or a multiple of 7 is $\dfrac{N(A \cup B)}{1000} = \dfrac{357}{1000} = 35.7\%$.

    c. By the difference rule the number of integers from 1 through 1000 that are neither multiples of 4 nor multiples of 7 is $1000 - 357 = 643$.

27. Let **A**, **B**, and **C** be the sets of all people who reported relief from drugs A, B, and C respectively. Then $N(\mathbf{A}) = 21$, $N(\mathbf{B}) = 21$, $N(\mathbf{C}) = 31$, $N(\mathbf{A} \cap \mathbf{B}) = 9$, $N(\mathbf{A} \cap \mathbf{C}) = 14$, $N(\mathbf{B} \cap \mathbf{C}) = 15$, and $N(\mathbf{A} \cup \mathbf{B} \cup \mathbf{C}) = 41$.

a. $N((\mathbf{A} \cup \mathbf{B} \cup \mathbf{C})^c) = 50 - N(\mathbf{A} \cup \mathbf{B} \cup \mathbf{C}) = 50 - 41 = 9$

b. $N(\mathbf{A} \cap \mathbf{B} \cap \mathbf{C}) = N(\mathbf{A} \cup \mathbf{B} \cup \mathbf{C}) - N(\mathbf{A}) - N(\mathbf{B}) - N(\mathbf{C}) + N(\mathbf{A} \cap \mathbf{B}) + N(\mathbf{A} \cap \mathbf{C}) + N(\mathbf{B} \cap \mathbf{C}) = 41 - 21 - 21 - 31 + 9 + 14 + 15 = 6$

c.

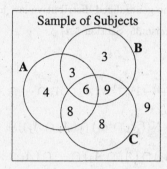

d. *Solution 1*: From the picture in part (c), it is clear that the number who got relief from A only is 4.

Solution 2: Because $(\mathbf{A} \cap \mathbf{B}) \cup (\mathbf{A} \cap \mathbf{C}) = \mathbf{A}$ and $(\mathbf{A} \cap \mathbf{B}) \cap (\mathbf{A} \cap \mathbf{C}) = \mathbf{A} \cap \mathbf{B} \cap \mathbf{C}$, we may apply the inclusion/exclusion rule to **A** to obtain the result that the number who got relief from A only equals $N(\mathbf{A}) - N(\mathbf{A} \cap \mathbf{B}) - N(\mathbf{A} \cap \mathbf{C}) + N(\mathbf{A} \cap \mathbf{B} \cap \mathbf{C}) = 21 - 9 - 14 + 6 = 4$.

33. *Proof (by mathematical induction)*: Let $P(k)$ be the property "If a finite set A equals the union of k distinct mutually disjoint subsets $A_1, A_2, \ldots, A_k$, then $N(A) = N(A_1) + N(A_2) + \cdots + N(A_k)$."

Show that the property is true for $k = 1$: Suppose a finite set A equals the "union" of one subset A_1, then $A = A_1$, and so $N(A) = N(A_1)$.

Show that for all integers $i \geq 1$, if the property is true for $k = i$ then it is true for $k = i + 1$: Let i be an integer with $i \geq 1$ and suppose the property is true for $n = i$. *[This is the inductive hypothesis.]* Let A be a finite set that equals the union of $i + 1$ distinct mutually disjoint subsets $A_1, A_2, \ldots, A_{i+1}$. Then $A = A_1 \cup A_2 \cup \cdots \cup A_{i+1}$ and $A_i \cap A_j = \emptyset$ for all integers i and j with $i \neq j$. Let B be the set $A_1 \cup A_2 \cup \cdots \cup A_i$. Then $A = B \cup A_{i+1}$ and $B \cap A_{i+1} = \emptyset$. *[For if $x \in B \cap A_{i+1}$, then $x \in A_1 \cup A_2 \cup \cdots \cup A_i$ and $x \in A_{i+1}$, which implies that $x \in A_j$, for some j with $1 \leq j \leq i$, and $x \in A_{i+1}$. But A_j and A_i are disjoint. Thus no such x exists.]* Hence A is the union of the two mutually disjoint sets B and A_{i+1}. Since B and A_{i+1} have no elements in common, the total number of elements in $B \cup A_{i+1}$ can be obtained by first counting the elements in B, next counting the elements in A_{i+1}, and then adding the two numbers together. It follows that $N(B \cup A_{i+1}) = N(B) + N(A_{i+1})$ which equals $N(A_1) + N(A_2) + \cdots + N(A_i) + N(A_{i+1})$ by inductive hypothesis. Hence $P(i + 1)$ is true *[as was to be shown]*.

36. *Proof (by mathematical induction)*:

Show that the property is true for $n = 2$: This was proved in one way in the text preceding Theorem 6.3.3 and in another way in the solution to exercise 34.

Show that for all integers $r \geq 2$, if the property is true for $n = r$ then it is true for $n = r + 1$: Let r be an integer with $r \geq 2$, and suppose that the formula holds for any collection of r finite sets. *[This is the inductive hypothesis.]* Let $A_1, A_2, \ldots, A_{r+1}$ be finite sets. Then

$N(A_1 \cup A_2 \cup \cdots \cup A_{r+1})$

$= N(A_1 \cup (A_2 \cup A_3 \cup \cdots \cup A_{r+1}))$ by the associative law for $\cup$

$= N(A_1) + N(A_2 \cup A_3 \cup \cdots \cup A_{r+1}) - N(A_1 \cap (A_2 \cup A_3 \cup \cdots \cup A_{r+1}))$

by the inclusion/exclusion rule for two sets

$= N(A_1) + N(A_2 \cup A_3 \cup \cdots \cup A_{r+1}) - N((A_1 \cap A_2) \cup (A_1 \cap A_3) \cup \cdots \cup (A_1 \cap A_{r+1}))$

by the generalized distributive law for sets

(exercise 35, Section 5.2)

$= N(A_1) + \Big(\sum_{2 \le i \le r+1} N(A_i) - \sum_{2 \le i < j \le r+1} N(A_i \cap A_j)$

$\qquad + \sum_{2 \le i < j < k \le r+1} N(A_i \cap A_j \cap A_k) - \cdots + (-1)^{r+1} N(A_2 \cap A_3 \cap \cdots \cap A_{r+1}) \Big)$

$\qquad - \Big(\sum_{2 \le i \le r+1} N(A_1 \cap A_i) - \sum_{2 \le i < j \le r+1} N((A_1 \cap A_i) \cap (A_1 \cap A_j)) + \cdots$

$\qquad\qquad + (-1)^{r+1} N((A_1 \cap A_2) \cap (A_1 \cap A_3) \cap \cdots \cap (A_1 \cap A_{r+1})) \Big)$

by inductive hypothesis

$= N(A_1) + \Big(\sum_{2 \le i \le r+1} N(A_i) - \sum_{2 \le i < j \le r+1} N(A_i \cap A_j)$

$\qquad + \sum_{2 \le i < j < k \le r+1} N(A_i \cap A_j \cap A_k) - \cdots + (-1)^{r+1} N(A_2 \cap A_3 \cap \cdots \cap A_{r+1}) \Big)$

$\qquad - \Big(\sum_{2 \le i \le r+1} N(A_1 \cap A_i) - \sum_{2 \le i < j \le r+1} N(A_1 \cap A_i \cap A_j) + \cdots$

$\qquad\qquad + (-1)^{r+1} N(A_1 \cap A_2 \cap A_3 \cap \cdots \cap A_{r+1}) \Big)$

$= \sum_{1 \le i \le r+1} N(A_i) - \sum_{1 \le i < j \le r+1} N(A_i \cap A_j) + \sum_{1 \le i < j < k \le r+1} N(A_i \cap A_j \cap A_k)$

$\qquad - \cdots + (-1)^{r+2} N(A_1 \cap A_3 \cap \cdots \cap A_{r+1}).$

[This is what was to be proved.]

Section 6.4

9. *a.* $\binom{40}{6} = 3,838,380$

12. The sum of two integers is even if, and only if, either both integers are even or both are odd *[see Example 3.2.3]*. Because $2 = 2 \cdot 1$ and $100 = 2 \cdot 50$, there are 50 even integers and thus 51 odd integers from 1 to 101 inclusive. Hence the number of distinct pairs is $\binom{50}{2} + \binom{51}{2} = 1225 + 1275 = 2500.$

18. $\binom{11}{1}\binom{10}{4}\binom{6}{4}\binom{2}{2} = \dfrac{11!}{1! \cdot 10!} \cdot \dfrac{10!}{4! \cdot 6!} \cdot \dfrac{6!}{4! \cdot 2!} \cdot \dfrac{2!}{2! \cdot 0!} = \dfrac{11!}{1! \cdot 4! \cdot 4! \cdot 2!} = 34,650$ *[which agrees with the result in Example 6.4.11]*

21. *c.* There are as many strings of length 4 with 2 a's and 2 b's as there are ways to choose 2 positions out of 4 into which to place the a's. Thus the answer is $\binom{4}{2} = \dfrac{4 \cdot 3}{2 \cdot 1} = 6.$

[Alternatively, one could think of choosing 2 positions (out of 4) for the a's and then 2 positions (out of the remaining 2) for the b's, and write the answer as $\binom{4}{2}\binom{2}{2} = 6$.]

27. Given nonnegative integers r and n with $r \leq n$, $P(n, r)$ is the set of r-permutations that can be formed from a set of n elements. Partition this set of r-permutations into subsets so that all the r-permutations in each subset are permutations of the same collection of elements. For instance, if $n = 5$ and $r = 3$ and $X = \{a, b, c, d, e\}$ is a set of $n = 5$ elements, then *acd, adc, cda, cad, dac,* and *dca* are all permutations of the same collection of elements, namely $\{a, c, d\}$. Thus each subset of the partition corresponds to a subset of X of size r. Furthermore, all subsets of the partition have the same size, namely $r!$ *[because there are $r!$ permutations of a set of r elements]*. Hence the number of subsets of X of size r equals the number of subsets of the partition. By the division rule, this equals the number of elements in the partition divided by the number of elements in each set of the partition, or $\dfrac{P(n, r)}{r!}$.

Section 6.5

6. $\dbinom{5 + n - 1}{5} = \dbinom{n + 4}{5} = \dfrac{(n + 4)(n + 3)(n + 2)(n + 1)n}{120}$

9. The number of iterations of the inner loop is the same as the number of integer triples (i, j, k) where $1 \leq k \leq j \leq i \leq n$. By reasoning similar to that of Example 6.5.3, the number of such triples is $\dbinom{n + 2}{3} = \dfrac{n(n + 1)(n + 2)}{6}$.

12. Think of the number 30 as divided into 30 individual units and the variables (y_1, y_2, y_3, y_4) as four categories into which these units are placed. The number of units in category y_i indicates the value of y_i in a solution of the equation. By Theorem 6.5.1, the number of ways to place 30 objects into four categories is $\dbinom{30 + 4 - 1}{30} = \dbinom{33}{30} = 5456$. So there are 5456 nonnegative integral solutions of the equation.

15. *a.* Think of the 30 kinds of balloons as the n categories and the 12 balloons to be chosen as the r objects. Each choice of 12 balloons is represented by a string of $30 - 1 = 29$ vertical bars (to separate the categories) and 12 crosses (to represent the chosen balloons). The total number of choices of 12 balloons of the 30 different kinds is the number of strings of 41 symbols (29 vertical bars and 12 crosses), namely, $\dbinom{12 + 30 - 1}{12} = \dbinom{41}{12} = 7,898,654,920$.

b. By the same reasoning as in part (a), the total number of choices of 50 balloons of the 30 different kinds is the number of strings that consist of $30 - 1 = 29$ vertical bars (to separate the categories) and 50 crosses (to represent the balloons that are chosen), namely, $\dbinom{50 + 30 - 1}{50} = \dbinom{79}{50} \cong 3.326779701 \times 10^{21}$. If at least one balloon of each kind is chosen, we can imagine choosing those 30 first and then choosing $50 - 30 = 20$ additional balloons. Again, by the same reasoning as in part (a), the number of ways to do this is the number of strings that consist of 29 vertical bars and 20 crosses, namely, $\dbinom{20 + 30 - 1}{20} = \dbinom{49}{20} \cong 2.827752735 \times 10^{13}$. Thus the probability that a combination of 50 balloons will contain at least one balloon of each kind is $\dfrac{\dbinom{(50 - 30) + 30 - 1}{50 - 30}}{\dbinom{50 + 30 - 1}{50}} = \dfrac{\dbinom{49}{20}}{\dbinom{79}{50}} \cong 8.5 \times 10^{-9}$.

Section 6.6

12.

$$\binom{n+3}{r} = \binom{n+2}{r-1} + \binom{n+2}{r}$$

$$= \left(\binom{n+1}{r-2} + \binom{n+1}{r-1}\right) + \left(\binom{n+1}{r-1} + \binom{n+1}{r}\right)$$

$$= \binom{n+1}{r-2} + 2 \cdot \binom{n+1}{r-1} + \binom{n+1}{r}$$

$$= \left(\binom{n}{r-3} + \binom{n}{r-2}\right) + 2 \cdot \left(\binom{n}{r-2} + \binom{n}{r-1}\right)$$

$$+ \left(\binom{n}{r-1} + \binom{n}{r}\right)$$

$$= \binom{n}{r-3} + 3 \cdot \binom{n}{r-2} + 3 \cdot \binom{n}{r-1} + \binom{n}{r}$$

15. *Proof:* Let n be an integer with $n \geq 1$. By exercise 14, $\sum_{i=1}^{n+1} \binom{i}{2} = \binom{n+2}{3}$. But for each

$i = 2, 3, \ldots, n+1$,

$$\binom{i}{2} = \frac{i!}{i! \cdot (i-2)!} = \frac{i \cdot (i-1)}{2} = \frac{(i-1) \cdot i}{2}.$$

So

$$\binom{n+2}{3} = \sum_{i=2}^{n+1} \binom{i}{2}$$

$$= \frac{1 \cdot 2}{2} + \frac{2 \cdot 3}{2} + \frac{3 \cdot 4}{2} + \cdots + \frac{n(n+1)}{2}$$

$$= \frac{1 \cdot 2 + 2 \cdot 3 + 3 \cdot 4 + \cdots + n(n+1)}{2}.$$

Multiplying both sides by 2 gives

$$1 \cdot 2 + 2 \cdot 3 + 3 \cdot 4 + \cdots + n(n+1) = 2\binom{n+2}{3}$$

[as was to be shown].

18. Suppose m and n are positive integers and r is a nonnegative integer that is less than or equal to the minimum value of m and n. Let S be a set of $m + n$ elements, and write $S = A \cup B$, where A is a subset of S with m elements, B is a subset of S with n elements, and $A \cap B = \emptyset$. The collection of subsets of r elements chosen from S can be partitioned as follows: those consisting of 0 elements chosen from A and r elements chosen from B, those consisting of 1 element chosen from A and $r - 1$ elements chosen from B, those consisting of 2 elements chosen from A and $r - 2$ elements chosen from B, and so forth, up to those consisting of r elements chosen from A and 0 elements chosen from B. By the multiplication rule, there are $\binom{m}{i}\binom{n}{r-i}$ ways to choose i objects from m and $r - i$ objects from n, and so the number of subsets of size r in which i elements are from A and $r - i$ objects are from B is $\binom{m}{i}\binom{n}{r-i}$.

Hence by the addition rule, the total number of subsets of S of size r is

$$\binom{m}{0}\binom{n}{r} + \binom{m}{1}\binom{n}{r-1} + \binom{m}{2}\binom{n}{r-2} + \cdots + \binom{m}{r}\binom{n}{0}.$$

But also since S has $m + n$ elements, the number of subsets of size r of S is $\binom{m+n}{r}$. So

$$\binom{m+n}{r} = \binom{m}{0}\binom{n}{r} + \binom{m}{1}\binom{n}{r-1} + \binom{m}{2}\binom{n}{r-2} + \cdots + \binom{m}{r}\binom{n}{0}.$$

21. *Proof*: Let p be a prime number and r an integer with $0 < r < p$. Then $\binom{p}{r} = \dfrac{p!}{r!(p-r)!} = \dfrac{p(p-1)!}{r!(p-r)!}$, or, equivalently, $p(p-1)! = \binom{p}{r}(r!(p-r)!)$. Now $\binom{p}{r}$ is an integer because it equals the number of subsets of size r that can be formed from a set with p elements. Thus we can apply the unique factorization theorem to express each side of this equation as a product of prime numbers. Clearly, p is a factor of the left-hand side, and so p must also be a factor of the right-hand side. But $0 < r < p$, and so p does not appear as one of the prime factors in either $r!$ or $(p-r)!$. Therefore, p must occur as one of the prime factors of $\binom{p}{r}$, and hence $\binom{p}{r}$ is divisible by p.

Section 6.7

6. *Solution 1*: $(u^2 - 3v)^4 = \binom{4}{0}(u^2)^4(-3v)^0 + \binom{4}{1}(u^2)^3(-3v)^1 + \binom{4}{2}(u^2)^2(-3v)^2$

$$+ \binom{4}{3}(u^2)^1(-3v)^3 + \binom{4}{4}(u^2)^0(-3v)^4$$

$$= u^8 - 12u^6v + 54u^4v^2 - 108u^2v^3 + 81v^4$$

Solution 2: An alternative solution is to first expand and simplify the expression $(a+b)^4$ and then substitute u^2 in place of a and $(-3v)$ in place of b and further simplify the result. Using this approach, we first apply the binomial theorem with $n = 4$ to obtain

$$(a+b)^4 = \binom{4}{0}a^4b^0 + \binom{4}{1}a^3b^1 + \binom{4}{2}a^2b^2 + \binom{4}{3}a^1b^3 + \binom{4}{4}b^4$$

$$= a^4 + 4a^3b + 6a^2b^2 + 4ab^3 + b^4.$$

Substituting u^2 in place of a and $(-3v)$ in place of b gives

$$(u^2 - 3v)^4 = (u^2 + (-3v))^4 = (u^2)^4 + 4(u^2)^3(-3v) + 6(u^2)^2(-3v)^2 + 4(u^2)(-3v)^3 + (-3v)^4$$

$$= u^8 - 12u^6v + 54u^4v^2 - 108u^2v^3 + 81v^4.$$

9. $\left(x^2 - \dfrac{1}{x}\right)^5$

$$= (x^2)^5 + \binom{5}{1}(x^2)^4\left(-\frac{1}{x}\right)^1 + \binom{5}{2}(x^2)^3\left(-\frac{1}{x}\right)^2 + \binom{5}{3}(x^2)^2\left(-\frac{1}{x}\right)^3$$

$$+ \binom{5}{4}(x^2)^1\left(-\frac{1}{x}\right)^4 + \left(-\frac{1}{x}\right)^5$$

$$= x^{10} - 5x^7 + 10x^4 - 10x + \frac{5}{x^2} - \frac{1}{x^5}$$

12. Term is $\binom{10}{3}(2x)^7 3^3$. Coefficient is $\dfrac{10!}{3! \cdot 7!} \cdot 2^7 \cdot 3^3 = 120 \cdot 128 \cdot 27 = 414{,}720.$

18. *Proof*: Let n be an integer with $n \geq 0$. Apply the binomial theorem with $a = 5$ and $b = -1$ to obtain

$$
\begin{aligned}
3^n &= (1+2)^n \\
&= \binom{n}{0}1^n 2^0 + \binom{n}{1}1^{n-1}2^1 + \cdots + \binom{n}{k}1^{n-k}2^k + \cdots + \binom{n}{n}1^{n-n}2^n \\
&= \binom{n}{0} + 2\binom{n}{1} + 2^2\binom{n}{2} + \cdots + 2^n\binom{n}{n}
\end{aligned}
$$

because $2^0 = 1$ and $1^{n-k} = 1$ for all integers k.

21. *Proof*: Let n be an integer with $n \geq 0$ and suppose x is any nonnegative real number. Apply the binomial theorem with $a = 1$ and $b = x$ to obtain

$$
\begin{aligned}
(1+x)^n &= \binom{n}{0}1^n x^0 + \binom{n}{1}1^{n-1}x^1 + \cdots + \binom{n}{k}1^{n-k}x^k + \cdots + \binom{n}{n}1^{n-n}x^n \\
&= \binom{n}{0} + \binom{n}{1}x + \binom{n}{2}x^2 + \cdots + \binom{n}{n}x^n \quad \text{because 1 raised to any power is 1} \\
&= 1 + nx + \frac{n(n-1)}{2}x^2 + \cdots + x^n.
\end{aligned}
$$

But each term to the right of nx is nonnegative. Hence $(1+x)^n \geq 1 + nx$.

27. Let m be an integer with $m \geq 0$. Then

$$
\sum_{k=0}^{m} \binom{m}{k} 2^{m-k} x^k = (2+x)^m \quad \text{by the binomial theorem with } a = 2 \text{ and } b = x.
$$

30. Let m be an integer with $m \geq 0$. Then

$$
\begin{aligned}
\sum_{i=0}^{m} \binom{m}{i} p^{m-i} q^{2i} &= \sum_{i=0}^{m} \binom{m}{i} p^{m-i}(q^2)^i \quad \text{by the laws of exponents} \\
&= (p+q^2)^m \quad \text{by the binomial theorem with } a = p \text{ and } b = q^2.
\end{aligned}
$$

33. Let n be an integer with $n \geq 0$. Then

$$
\begin{aligned}
\sum_{k=0}^{n} \binom{n}{k} 3^{2n-2k} 2^{2k} &= \sum_{k=0}^{n} \binom{n}{k} (3^2)^{n-k}(2^2)^k \\
&\qquad \text{by the laws of exponents} \\
&= \sum_{k=0}^{n} \binom{n}{k} 9^{n-k} 4^k \\
&\qquad \text{because } 3^2 = 9 \text{ and } 2^2 = 4 \\
&= (9+4)^n \\
&\qquad \text{by the binomial theorem with } a = 9 \text{ and } b = -4 \\
&= 13^n.
\end{aligned}
$$

36. *a.* Let n be an integer with $n \geq 0$. Apply the binomial theorem with $a = 1$ and $b = x$ to obtain

$$
(1+x)^n = \sum_{k=0}^{n} \binom{n}{k} 1^{n-k} x^k = \sum_{k=0}^{n} \binom{n}{k} x^k \quad \text{because any power of 1 is 1.}
$$

c. (ii) Let n be an integer with $n \geq 1$. Apply the formula from part (b) with $x = -1$ to obtain

$$
0 = n(1 + (-1))^{n-1} = \sum_{k=1}^{n} \binom{n}{k} k(-1)^{k-1}.
$$

d. Apply the formula of part (b) with $x = 3$ to obtain

$$
n \cdot 4^{n-1} = n(1+3)^{n-1} = \sum_{k=1}^{n} \binom{n}{k} k 3^{k-1} = \sum_{k=1}^{n} \binom{n}{k} k 3^k 3^{-1} = \frac{1}{3}\sum_{k=1}^{n} \binom{n}{k} k 3^k.
$$

So $\displaystyle\sum_{k=1}^{n}\binom{n}{k}k3^k = 3n4^{n-1}$.

Section 6.8

3. *a.* $P(A \cup B) = 0.4 + 0.2 = 0.6$

   b. By the formula for the probability of a general union and because $S = A \cup B \cup C$,
   $$P(S) = ((A \cup B) \cup C) = P(A \cup B) + P(C) - P((A \cup B) \cap C).$$
   Suppose $P(C) = 0.2$. Then, since $P(S) = 1$,
   $$1 = 0.6 + 0.2 - P((A \cup B) \cap C) = 0.8 - P((A \cup B) \cap C).$$
   Solving for $P((A \cup B) \cap C)$ gives $P((A \cup B) \cap C) = -0.2$, which is impossible. Hence $P(C) \neq 0.2$.

6. First note that we can apply the formula for the probability of the complement of an event to obtain $0.3 = P(U^c) = 1 - P(U)$. Solving for $P(U)$ gives $P(U) = 0.7$. Second, observe that by De Morgan's law $U^c \cup V^c = (U \cap V)^c$. Thus $0.4 = P(U^c \cup V^c) = P((U \cap V)^c) = 1 - P(U \cap V)$. Solving for $P(U \cap V)$ gives $P(U \cap V) = 0.6$. So, by the formula for the union of two events, $P(U \cup V) = P(U) + P(V) - P(U \cap V) = 0.7 + 0.6 - 0.6 = 0.7$.

9. *b.* By part (a), $P(A \cup B) = 0.7$. So, since $C = (A \cup B)^c$, by the formula for the probability of the complement of an event, $P(C) = 1 - P(A \cup B) = 1 - 0.7 = 0.3$.

   c. By the formula for the probability of the complement of an event, $P(A^c) = 1 - P(A) = 1 - 0.4 = 0.6$.

   e. By De Morgan's law $A^c \cup B^c = (A \cap B)^c$. Thus, the formula for the probability of the complement of an event, $P(A^c \cup B^c) = P((A \cap B)^c) = 1 - P(A \cap B) = 1 - 0.2 = 0.8$.

   f. Solution 1: Because $C = S - (A \cup B)$, $C = (A \cup B)^c$. Thus by substitution, De Morgan's law, and the associative, commutative, and idempotent properties of $\cap$,
   $$B^c \cap C = B^c \cap (A \cup B)^c = B^c \cap (A^c \cap B^c) = (B^c \cap A^c) \cap B^c = (A^c \cap B^c) \cap B^c = A^c \cap (B^c \cap B^c) = A^c \cap B^c = (A \cup B)^c = C.$$ Hence, by part (b),
   $$P(B^c \cap C) = P(C) = 0.3.$$

   Solution 2: Because $C = S - (A \cup B)$, $C = (A \cup B)^c$. Thus by De Morgan's law, $C = A^c \cap B^c$. Now $A^c \cap B^c \subseteq B^c$ (by Theorem 5.2.1(1)a) and hence $B^c \cap C = C$ (by Theorem 5.2.3a). Therefore $P(B^c \cap C) = P(C) = 0.3$.

12. *Proof 1*: Suppose S is any sample space and U and V are any events in S. First note that by the set difference, distributive, universal bound, and identity laws, $(V \cap U) \cup (V - U) = (V \cap U) \cup (V \cap U^c) = V \cap (U \cup U^c) = V \cap S = V$. Next, observe that if $x \in (V \cap U) \cap (V - U)$, then, by definition of intersection, $x \in (V \cap U)$ and $x \in (V - U)$, and so, by definition of intersection and set difference, $x \in V$, $x \in U$, $x \in V$, and $x \notin U$, and hence, in particular, $x \in U$ and $x \notin U$, which is impossible. It follows that $(V \cap U) \cap (V - U) = \emptyset$. Thus, by substitution and by probability axiom 3 (the formula for the probability of mutually disjoint events), $P(V) = P((V \cap U) \cup (V - U)) = P(V \cap U) + P(V - U)$. Solving for $P(V - U)$ gives $P(V - U) = P(V) - P(U \cap V)$.

   Proof 2: Suppose S is any sample space and U and V are any events in S. First note that by the set difference, distributive, universal bound, and identity laws, $U \cup (V - U) = U \cup (V \cap U^c) = (U \cup V) \cap (U \cup U^c) = (U \cup V) \cap S = U \cup V$. Next, observe that, by the same sequence of steps as in the solution to exercise 11, $U \cap (V - U) = \emptyset$. Thus, by substitution, $P(U \cup V) = P(U \cup (V - U)) = P(U) + P(V - U)$. But also by the formula for the probability of a general union, $P(U \cup V) = P(U) + P(V) - P(U \cap V)$. Equating the two expressions for $P(U \cup V)$ gives $P(U) + P(V - U) = P(U) + P(V) - P(U \cap V)$. Subtracting $P(U)$ from both sides gives $P(V - U) = P(V) - P(U \cap V)$.

15. *Solution 1*: The net gain for the first prize winner is $\$10,000,000 - \$0.60 = \$9,999,999.40$, that for the second prize winner is $\$1,000,000 - \$0.60 = \$999,999.40$, and that for the third prize winner is $\$50,000 - \$0.60 = 49,999.40$. Each of the other 29,999,997 million people who mail back an entry form has a net loss of $\$0.60$. Because all of the 30 million entry forms have an equal chance of winning the prizes, the expected gain or loss is

$$\$9999999.40 \cdot \frac{1}{30000000} + \$999999.40 \cdot \frac{1}{30000000} + \$49999.40 \cdot \frac{1}{30000000} - (\$0.60) \cdot \frac{29999997}{30000000}$$

$$\cong -\$0.23,$$

or an expected loss of about 23 cents per person.

Solution 2: The total amount spent by the 30 million people who return entry forms is $30,000,000 \cdot \$0.60 = \$18,000,000$. The total amount of prize money awarded is $\$10,000,000 + \$1,000,000 + \$50,000 = \$11,050,000$. Thus the net loss is $\$18,000,000 - \$11,050,000 = \$6,950,000$, and so the expected loss per person is $6950000/30000000 \cong -\$0.23$, or about 23 cents per person.

18. Let 2_1 and 2_2 denote the two balls with the number 2, let 8_1 and 8_2 denote the two balls with the number 8, and let 1 denote the other ball. There are $\binom{5}{3} = 10$ subsets of 3 balls that can be chosen from the urn. The following table shows the sums of the numbers on the balls in each set and the corresponding probabilities:

Subset	Sum s	Probability of s
$\{1, 2_1, 2_2\}$	5	1/10
$\{1, 2_1, 8_1\}, \{1, 2_2, 8_1\}, \{1, 2_1, 8_2\}, \{1, 2_2, 8_2\}$	11	4/10
$\{2_1, 2_2, 8_1\}, \{2_1, 2_2, 8_2\}$	12	2/10
$\{1, 8_1, 8_2\}$	17	1/10
$\{2_1, 8_1, 8_2\}, \{2_2, 8_1, 8_2\}$	18	2/10

Thus the expected value is $5 \cdot \frac{1}{10} + 11 \cdot \frac{4}{10} + 12 \cdot \frac{2}{10} + 17 \cdot \frac{1}{10} + 18 \cdot \frac{2}{10} = \frac{126}{10} = 12.6$.

21. When a coin is tossed 4 times, there are $2^4 = 16$ possible outcomes and there are $\binom{4}{h}$ ways to obtain exactly h heads (as shown by the technique illustrated in Example 6.4.10). The following table shows the possible outcomes of the tosses, the amount gained or lost for each outcome, the number of ways the outcomes can occur, and the probabilities of the outcomes.

Number of Heads	Net Gain (or Loss)	Number of Ways	Probability
0	−$3	$\binom{4}{0} = 1$	1/16
1	−$2	$\binom{4}{1} = 4$	4/16
2	−$1	$\binom{4}{2} = 6$	6/16
3	$2	$\binom{4}{3} = 4$	4/16
4	$3	$\binom{4}{4} = 1$	1/16

Thus the expected value is $(-\$3) \cdot \frac{1}{16} + (-\$2) \cdot \frac{4}{16} + (-\$1) \cdot \frac{6}{16} + \$2 \cdot \frac{4}{16} + \$3 \cdot \frac{1}{16} = -\$\frac{6}{16} = -\$0.375$. So this game has an expected loss of 37.5 cents.

Section 6.9

3. *b.* Let A be the event that a randomly chosen person tests positive for a condition, let B_1 be the event that the person has the condition, and let B_2 be the event that the person does not have the condition. Then the probability that a randomly chosen person tests negative for the condition is A^c. Then the probability of a false positive is $P(A \mid B_2)$, the probability of a false negative is $P(A^c \mid B_1)$, the probability that a person who actually has the condition tests positive for it is $P(A \mid B_1)$, and the probability that a person who does not have the condition tests negative for it is $P(A^c \mid B_2)$.

(1) Suppose the probability of a false positive is 4%. This means that $P(A \mid B_2) = 4\% = 0.04$. By part (a), $P(A^c \mid B_2) = 1 - P(A \mid B_2) = 1 - 0.04 = 0.96$, and so the probability that a person who does not have the condition tests negative for it is 96%.

(2) Suppose the probability of a false negative is 1%. This means that $P(A^c \mid B_1) = 1\% = 0.01$. By part (a), $P(A^c \mid B_1) = 1 - P(A \mid B_1)$, and so $P(A \mid B_1) = 1 - P(A^c \mid B_1) = 1 - 0.01 = 0.99$. Thus the probability that a person who has the condition tests positive for it is 99%.

6. Let R_1 be the event that the first ball is red, R_2 the event that the second ball is red, B_1 the event that the first ball is blue, and B_2 the event that the second ball is blue. Then $P(R_1) = \dfrac{30}{70}$, $P(B_1) = \dfrac{40}{70}$, $P(R_2 \mid R_1) = \dfrac{29}{69}$, $P(R_2 \mid B_1) = \dfrac{30}{69}$, and $P(B_2 \mid B_1) = \dfrac{39}{69}$.

a. The probability that both balls are red is $P(R_1 \cap R_2) = P(R_2 \mid R_1)P(R_1) = \dfrac{29}{69} \cdot \dfrac{30}{70} = \dfrac{29}{161} \cong 18.0\%$.

b. The probability that the second ball is red but the first ball is not is $P(B_1 \cap R_2) = P(R_2 \mid B_1)P(B_1) = \dfrac{30}{69} \cdot \dfrac{40}{70} = \dfrac{40}{161} \cong 24.8\%$.

c. Because $B_1 \cap R_1 = \emptyset$ and $B_1 \cup R_1$ is the entire sample space S, $R_2 = S \cap R_2 = (B_1 \cap R_2) \cup (R_1 \cap R_2)$ and $(B_1 \cap R_2) \cap (R_1 \cap R_2) = \emptyset$. Thus the probability that the second ball is red is $P(R_2) = P((B_1 \cap R_2) \cup (R_1 \cap R_2)) = P(B_1 \cap R_2) + P(R_1 \cap R_2) = \dfrac{40}{161} + \dfrac{29}{161} = \dfrac{69}{161} \cong 18\% + 25\% = 42.8\%$.

d. Solution 1: The probability that at least one of the balls is red is $P(R_1 \cup R_2) = P(R_1) + P(R_2) - P(R_1 \cap R_2) = \dfrac{30}{70} + \dfrac{69}{161} - \dfrac{29}{161} \cong 67.7\%$.

Solution 2: The event that at least one of the balls is red is the complement of the event that both balls are blue. Thus $P(R_1 \cup R_2) = 1 - P(B_1 \cap B_2) = 1 - P(B_2 \mid B_1)P(B_1) = 1 - \dfrac{39}{69} \cdot \dfrac{40}{70} \cong 67.7\%$

9. *Proof*: Let S be a sample space, let A be an event in S with $P(A) \neq 0$, and suppose that $B_1, B_2, \ldots, B_n$ are mutually disjoint events in S such that $S = B_1 \cup B_2 \cup \cdots \cup B_n$. By the generalized distributive law for sets, $A = A \cap S = A \cap (B_1 \cup B_2 \cup \cdots \cup B_n) = (A \cap B_1) \cup (A \cap B_2) \cup \cdots \cup (A \cap B_n)$. Also because $B_1, B_2, \ldots, B_n$ are mutually disjoint and by the associative, commutative, and universal bound laws for sets, $(A \cap B_i) \cap (A \cap B_j) = (A \cap A) \cap (B_i \cap B_j) = (A \cap A) \cap \emptyset = \emptyset$ for all integers i and j with $1 \leq i \leq n$, $1 \leq j \leq n$, and $i \neq j$. Thus the sets $(A \cap B_1), (A \cap B_2), \ldots, (A \cap B_n)$ are also mutually disjoint, and so, by substitution and the definition of conditional probability,

$$
\begin{aligned}
P(A) &= P((A \cap B_1) \cup (A \cap B_2) \cup \cdots \cup (A \cap B_n)) \\
&= P(A \cap B_1) + P(A \cap B_2) + \cdots + P(A \cap B_n) \\
&= P(A \mid B_1)P(B_1) + P(A \mid B_2)P(B_2) + \cdots + P(A \mid B_n)P(B_n).
\end{aligned}
$$

Also by definition of conditional probability,

$$P(A \mid B_k)P(B_k) = \frac{P(A \cap B_k)}{P(B_k)} P(B_k) = P(A \cap B_k).$$

Putting these results together gives

$$P(B_k \mid A) = \frac{P(A \cap B_k)}{P(A)} = \frac{P(A \mid B_k)P(B_k)}{P(A \mid B_1)P(B_1) + P(A \mid B_2)P(B_2) + \cdots + P(A \mid B_n)P(B_n)}$$

[as was to be shown].

12. *a.* Let B_1 be the event that the first urn is chosen, B_2 the event that the second urn is chosen, and A the event that the chosen ball is green. Then

$$P(B_1) = \frac{4}{10} = \frac{2}{5}, \qquad P(B_2) = \frac{6}{10} = \frac{3}{5}, \qquad P(A \mid B_1) = \frac{25}{35} \quad \text{and} \quad P(A \mid B_2) = \frac{15}{37}.$$

$$P(A \cap B_1) = P(A \mid B_1)P(B_1) = \frac{25}{35} \cdot \frac{2}{5} = \frac{2}{7}.$$

Also

$$P(A \cap B_2) = P(A \mid B_2)P(B_2) = \frac{15}{37} \cdot \frac{3}{5} = \frac{9}{37}.$$

Now A is the disjoint union of $A \cap B_1$ and $A \cap B_2$. So

$$P(A) = P(A \cap B_1) + P(A \cap B_2) = \frac{2}{7} + \frac{9}{37} = \frac{137}{259} \cong 52.9\%.$$

So the probability that the chosen ball is green is approximately 52.9%.

b. Solution 1 (using Bayes' theorem): Given that the chosen ball is green, the probability that it came from the first urn is $P(B_1 \mid A)$. By Bayes' theorem and the computations in part (a),

$$P(B_1 \mid A) = \frac{P(A \mid B_1)P(B_1)}{P(A \mid B_1)P(B_1) + P(A \mid B_2)P(B_2)} = \frac{\frac{2}{7}}{\frac{2}{7} + \frac{9}{37}} = \frac{74}{137} \cong 54.0\%.$$

Solution 2 (without explicit use of Bayes' theorem): Given that the chosen ball is green, the probability that it came from the first urn is $P(B_1 \mid A)$. By the results of part (a),

$$P(B_1 \mid A) = \frac{P(A \cap B_1)}{P(A)} = \frac{\frac{2}{7}}{\frac{137}{259}} = \frac{74}{137} \cong 54.0\%.$$

15. Let B_1 be the event that a randomly chosen piece of produce is from supplier X, B_2 the event that a randomly chosen piece of produce is from supplier Y, B_3 the event that a randomly chosen piece of produce is from supplier Z, and A the event that a randomly chosen piece of produce purchased from the store is superior grade. From the information given in the problem statement, we know that $P(B_1) = 20\% = 0.2$, $P(B_2) = 45\% = 0.45$, $P(B_3) = 35\% = 0.35$, $P(A \mid B_1) = 12\% = 0.12$, $P(A \mid B_2) = 8\% = 0.08$, and $P(A \mid B_3) = 15\% = 0.15$.

a. By definition of conditional probability,

$$P(A \cap B_1) = P(A \mid B_1)P(B_1) = 0.12 \cdot 0.2 = 0.024,$$
$$P(A \cap B_2) = P(A \mid B_2)P(B_2) = 0.08 \cdot 0.45 = 0.036,$$
$$P(A \cap B_3) = P(A \mid B_3)P(B_3) = 0.15 \cdot 0.35 = 0.0525.$$

Now because B_1, B_2, and B_3 are disjoint and because their union is the entire sample space, A is the disjoint union of $A \cap B_1$, $A \cap B_2$, and $A \cap B_3$. Thus

$$P(A) = P(A \cap B_1) + P(A \cap B_2) + P(A \cap B_3) = 0.024 + 0.036 + 0.0525 = 0.1125 = 11.25\%.$$

b. Solution 1 (by direct application of Bayes' theorem): Given that the chosen part is defective, the probability that it came from the first factory is $P(B_1 \mid A)$. By Bayes' theorem,

$$P(B_1 \mid A) = \frac{P(A \mid B_1)P(B_1)}{P(A \mid B_1)P(B_1) + P(A \mid B_2)P(B_2) + P(A \mid B_3)P(B_3)}$$

$$= \frac{0.12 \cdot 0.2}{0.12 \cdot 0.2 + 0.08 \cdot 0.45 + 0.15 \cdot 0.35}$$

$$\cong 21.3\%.$$

Solution 2 (without explicit use of Bayes' theorem): Given that the chosen ball is green, the probability that it came from the first urn is $P(B_1 \mid A)$. By definition of conditional probability and the results of part (a), $P(B_1 \mid A) = \dfrac{P(A \cap B_1)}{P(A)} = \dfrac{0.024}{0.1125} \cong 21.3\%$.

21. *Alternative proof to that given in Appendix B*: Suppose A and B are independent events in a sample space S. Then $P(A \cap B) = P(A)P(B)$. In case $P(B) = 0$, then, by the result of exercise 11, Section 6.8, $P(A^c \cap B) = 0$ (because $A^c \cap B \subseteq B$). Thus, $P(A^c \cap B) = 0 = P(A^c) \cdot 0 = P(A^c)P(B)$. In case $P(B) \neq 0$,

$$
\begin{aligned}
P(A^c \cap B) &= P(A^c \mid B)P(B) && \text{by formula 6.9.2} \\
&= [1 - P(A \mid B)]P(B) && \text{by exercise 3a} \\
&= [1 - P(A)]P(B) && \text{by the result of exercise 17} \\
&= P(A^c)P(B) && \text{by the formula for the complement of an event (6.8.1).}
\end{aligned}
$$

It follows by definition of independence that in either case A^c and B are independent.

24. Let A be the event that a randomly chosen error is missed by proofreader X, and let B be the event that the error is missed by proofreader Y. Then $P(A) = 0.12$ and $P(B) = 0.15$.

a. Because the proofreaders work independently, $P(A \cap B) = P(A)P(B)$. Hence the probability that the error is missed by both proofreaders is $P(A \cap B) = P(A)P)B) = 0.12 \cdot 0.15 = 0.018 = 1.8\%$.

27. *Solution*: The family could have two boys, two girls, or one boy and one girl. Let the subscript 1 denote the firstborn child (understanding that in the case of twins this might be by only a few moments), and let the subscript 2 denote the secondborn child. Then we can let (B_1G_2, B_1) denote the outcome that the firstborn child is a boy, the secondborn is a girl, and the child you meet is the boy. Similarly, we can let (B_1B_2, B_2) denote the outcome that both the firstborn and the secondborn are boys and the child you meet is the secondborn boy. When this notational scheme is used for the entire set of possible outcomes for the genders of the children and the gender of the child you meet, all outcomes are equally likely and the sample space is denoted by

$$\{(B_1B_2, B_1), (B_1B_2, B_2), (B_1G_2, B_1), (B_1G_2, G_2), (G_1B_2, G_1), (G_1B_2, B_2), (G_1G_2, G_1), (G_1G_2, G_2)\}.$$

The event that you meet one of the children and it is a boy is

$$\{(B_1B_2, B_1), (B_1B_2, B_2), (B_1G_2, B_1), (G_1B_2, B_2)\}.$$

The probability of this event is $4/8 = 1/2$.

Discussion: An intuitive way to see this conclusion is to realize that the fact that you happen to meet one of the children see that it is a boy gives you no information about the gender of the other child. Because each of the children is equally likely to be a boy, the probability that the other child is a boy is 1/2. Consider the following situation in which the probabilities are identical to the situation described in the exercise. A person tosses two fair coins and immediately covers them so that you cannot see which faces are up. The person then reveals one of the coins, and you see that it is heads. This action on the person's part has given you no information about the other coin; the probability that the other coin has also landed heads up is 1/2.

30. *a.* $P(0 \text{ false positives}) = \begin{bmatrix} \text{the number of ways} \\ 0 \text{ false positives} \\ \text{can be obtained} \\ \text{over a ten-year period} \end{bmatrix} \left(P\left(\begin{array}{c} \text{false} \\ \text{positive} \end{array} \right) \right)^0 \left(P\left(\begin{array}{c} \text{not a false} \\ \text{positive} \end{array} \right) \right)^{10}$

$$= \binom{10}{0} 0.96^{10} = 1 \cdot 0.96^{10} \cong 0.665 = 66.5\%$$

c. $P(2 \text{ false positives}) = \begin{bmatrix} \text{the number of ways 2} \\ \text{false positives can be} \\ \text{obtained over a} \\ \text{ten-year period} \end{bmatrix} \left(P\left(\begin{array}{c} \text{false} \\ \text{positive} \end{array} \right) \right)^2 \left(P\left(\begin{array}{c} \text{not a false} \\ \text{positive} \end{array} \right) \right)^8$

$$= \binom{10}{2} 0.04^2 \cdot 0.96^8 = 45 \cdot 0.04^2 \cdot 0.96^8 = 0.0594 \cong 5.2\%$$

d. Let T be the event that a woman's test result is positive one year, and let C be the event that the woman has breast cancer.

(i) By Bayes formula, the probability of C given T is

$$P(C|T) = \frac{P(T|C)P(C)}{P(T|C)P(C) + P(T|C^c)P(C^c)}$$

$$= \frac{(0.98)(0.0002)}{(0.98)(0.0002) + (0.04)(0.9998)}$$

$$\cong 0.00488 = 4.88\%.$$

(ii) The event that a woman's test result is negative one year is T^c. By Bayes formula, the probability of C given T^c is

$$P(C|T^c) = \frac{P(T^c|C)P(C)}{P(T^c|C)P(C) + P(T^c|C^c)P(C^c)}$$

$$= \frac{(0.02)(0.0002)}{(0.02)(0.0002) + (0.98)(0.9998)}$$

$$\cong 0.000004 = 0.0004\%.$$

General Review for Chapter 6

Probability

- What is the sample space of an experiment? *(p. 299)*
- What is an event in the sample space? *(p. 299)*
- What is the probability of an event when all the outcomes are equally likely? *(p. 299)*

Counting

- If m and n are integers with $m \leq n$, how many integers are there from m to n inclusive? *(p. 302)*
- How do you construct a possibility tree? *(p. 306)*
- What are the multiplication rule, the addition rule, and the difference rule? *(pp. 308, 321, 322)*
- When should you use the multiplication rule and when should you use the addition rule? *(p. 345)*
- What is the inclusion/exclusion formula? *(p. 327)*
- What is a permutation? an r-permutation? *(p. 313, 315)*
- What is $P(n,r)$? *(p. 315)*
- How does the multiplication rule give rise to $P(n,r)$? *(p. 315)*
- What is $\binom{n}{r}$? *(p. 334)*
- What is an r-combination? *(p. 334)*
- What formulas are used to compute $\binom{n}{r}$ by hand? *(p. 337)*
- What is an r-combination with repetition allowed (or a multiset of size r)? *(p. 349)*
- How many r-combinations with repetition allowed can be selected from a set of n elements? *(p. 351)*

Pascal's Formula and the Binomial Theorem

- What is Pascal's formula? Can you apply it in various situations? *(p. 358)*
- How is Pascal's formula proved? *(p. 360)*
- What is the binomial theorem? *(p. 364)*
- How is the binomial theorem proved? *(p. 364-367)*

Probability Axioms and Expected Value

- What is the range of values for the probability of an event? *(p. 370)*
- What is the probability of an entire sample space? *(p. 370)*
- What is the probability of the empty set? *(p. 370)*
- If A and B are disjoint events in a sample space S, what is $P(A \cup B)$? *(p. 370)*
- If A is an event in a sample space S, what is $P(A^c)$? *(p. 371)*
- If A and B are any events in a sample space S, what is $P(A \cup B)$? *(p. 371)*
- How do you compute the expected value of a random experiment or process, if the possible outcomes are all real numbers and you know the probability of each outcome? *(p. 373)*
- What is the conditional probability of one event given another event? *(p. 376)*
- What is Bayes' theorem? *(p. 379)*
- What does it mean for two events to be independent? *(p. 381)*

- What is the probability of an intersection of two independent events? *(p. 385)*
- What does it mean for events to be mutually independent? *(p. 384)*
- What is the probability of an intersection of mutually independent events? *(p. 385)*

Test Your Understanding: Chapter 6

Test yourself by filling in the blanks.

1. A sample space of a random process or experiment is _____.

2. An event in a sample space is _____.

3. To compute the probability of an event using the equally likely probability formula, you take the ratio of the _____ to the _____.

4. If $m \leq n$, the number of integers from m to n inclusive is _____.

5. The multiplication rule says that if an operation can be performed in k steps and, for each i with $1 \leq i \leq k$, the ith step can be performed in n_i ways (regardless of how previous steps were performed), then _____.

6. A permutation of a set of elements is _____.

7. The number of permutations of a set of n elements equals _____.

8. An r-permutation of a set of n elements is _____.

9. The number of r-permutations of a set of n elements is denoted _____.

10. One formula for the number of r-permutations of a set of n elements is _____ and another formula is _____.

11. The addition rule says that if a finite set A equals the union of k distinct mutually disjoint subsets $A_1, A_2, \ldots, A_k$, then _____.

12. The difference rule says that if A is a finite set and B is a subset of A, then _____.

13. If S is a finite sample space and A is an event in S, then the probability of A^c equals _____.

14. The inclusion/exclusion rule for two sets says that if A and B are any finite sets, then _____.

15. The inclusion/exclusion rule for three sets says that if A, B, and C are any finite sets, then _____.

16. The number of subsets of size r that can be formed from a set with n elements is denoted _____, which is read as _____.

17. Alternative phrases used to describe a subset of size r that is formed from a set with n elements are _____ and _____.

18. Two ordered selections are said to be the same if _____ and also if _____.

19. Two unordered selections are said to be the same if _____, regardless of _____.

20. The formula relating $\binom{n}{r}$ and $P(n, r)$ is _____.

21. Additional formulas for $\binom{n}{r}$ are _____ and _____.

22. The phrase "at least n" means ____, and the phrase "at most n" means ____.

23. Suppose a collection consists of n objects of which, for each i with $1 \le i \le k$, n_i are of type i and are indistinguishable from each other. Also suppose that $n = n_1 + n_2 + \cdots + n_k$. Then the number of distinct permutations of the n objects is ____.

24. Given a set $X = \{x_1, x_2, \ldots x_n\}$, an r-combination with repetition allowed, or a multiset of size r, chosen from X is ____, which is denoted ____.

25. If $X = \{x_1, x_2, \ldots x_n\}$, the number of r-combinations with repetition allowed (or multisets of size r) chosen from X is ____.

26. When choosing k elements from a set of n elements, order may or may not matter and repetition may or may not be allowed.

   - The number of ways to choose the k elements when repetition is allowed and order matters is ____.

   - The number of ways to choose the k elements when repetition is not allowed and order matters is ____.

   - The number of ways to choose the k elements when repetition is not allowed and order does not matter is ____.

   - The number of ways to choose the k elements when repetition is allowed and order does not matter is ____.

27. If n is a nonnegative integer, then $\binom{n}{n} = $ ____ and $\binom{n}{1} = $ ____.

28. If n and r are nonnegative integers with $r \le n$, then the relation between $\binom{n}{r}$ and $\binom{n}{n-r}$ is ____.

29. Pascal's formula says that if n and r are positive integers with $r \le n$, then ____.

30. The crux of the algebraic proof of Pascal's formula is that to add two fractions you need to express both of them with a ____.

31. The crux of the combinatorial proof of Pascal's formula is that the set of subsets of size r of a set $\{x_1, x_2, \ldots, x_n\}$ can be partitioned into the set of subsets of size r that contain ____ and those that ____.

32. The binomial theorem says that given any real numbers a and b and any nonnegative integer n, ____.

33. The crux of the algebraic proof of the binomial theorem is that, after making a change of variable so that two summations have the same lower and upper limits, you use the fact that $\binom{m}{k} + \binom{m}{k-1} = $ ____.

34. The crux of the combinatorial proof of the binomial theorem is that the number of ways to arrange k b's and $(n-k)$ a's in order is ____.

35. If A is an event in a sample space S, $P(A)$ can take values between ____ and ____. Moreover, $P(S) = $ ____, and $P(\emptyset) = $ ____.

36. If A and B are disjoint events in a sample space S, $P(A \cup B) = $ ____.

37. If A is an event in a sample space S, $P(A^c) = $ ____.

38. If A and B are any events in a sample space S, $P(A \cup B) =$ ____.

39. If the possible outcomes of a random process or experiment are real numbers $a_1, a_2, \ldots, a_n$, which occur with probabilities $p_1, p_2, \ldots, p_n$, then the expected value of the process is ____.

40. If A and B are any events in a sample space S and $P(A) \neq 0$, then the conditional probability of B given A is $P(B|A) =$ ____.

41. Bayes' theorem says that if a sample space S is a union of mutually disjoint events $B_1, B_2, \ldots, B_n$ with nonzero probabilities, if A is an event in S with $P(A) \neq 0$, and if k is an integer with $1 \leq k \leq n$, then ____.

42. Events A and B in a sample space S are independent if, and only if, ____.

43. Events A, B, and C in a sample space S are mutually independent if, and only if, ____, ____, ____, and ____.

Answers

1. the set of all outcomes of the random process or experiment
2. a subset of the sample space
3. number of outcomes in the event; total number of outcomes
4. $n - m + 1$
5. the operation as a whole can be performed in $n_1 n_2 \cdots n_k$ ways
6. an ordering of the elements of the set in a row
7. $n!$
8. an ordered selection of r of the elements of the set
9. $P(n, r)$
10. $n(n-1)(n-2) \cdots (n-r+1)$; $\dfrac{n!}{(n-r)!}$
11. the number of elements in A equals $N(A_1) + N(A_2) + \cdots + N(A_k)$
12. the number of elements in $A - B$ is the difference between the number of elements in A minus the number of elements in B
13. $1 - P(A)$
14. $N(A \cup B) = N(A) + N(B) - N(A \cap B)$
15. $N(A \cup B \cup C) = N(A) + N(B) + N(C) - N(A \cap B) - N(A \cap C) - N(B \cap C) + N(A \cap B \cap C)$
16. $\dbinom{n}{r}$; n choose r
17. an r-combination of the set of n elements; an unordered selection of r elements chosen from the set of n elements
18. the elements chosen are the same; the elements are chosen in the same order
19. the elements chosen are the same; the order in which the elements are chosen
20. $\dbinom{n}{r} = \dfrac{P(n, r)}{r!}$
21. $\dbinom{n}{r} = \dfrac{n(n-1)(n-2) \cdots (n-r+1)}{r!}$; $\dbinom{n}{r} = \dfrac{n!}{r!(n-r)!}$
22. n or more; n or fewer
23. $\dbinom{n}{n_1}\dbinom{n-n_1}{n_2}\dbinom{n-n_1-n_2}{n_3} \cdots \dbinom{n-n_1-n_2-\cdots-n_{k-1}}{n_k} = \dfrac{n!}{n_1! n_2! n_3! \cdots n_k!}$
24. an unordered selection of elements taken from X with repetition allowed $[x_1, x_2, \ldots x_{i_r}]$ where each x_{i_j} is in X and some of the x_{i_j} may equal each other

25. $\dbinom{k+n-1}{k}$

26. n^k; $n(n-1)(n-2)\cdots(n-k+1)$; $\dbinom{n}{k}$; $\dbinom{k+n-1}{k}$

27. 1; n

28. $\dbinom{n}{r} = \dbinom{n}{n-r}$

29. $\dbinom{n+1}{r} = \dbinom{n}{r-1} + \dbinom{n}{r}$

30. common denominator

31. x_n; do not contain x_n

32. $(a+b)^n = \displaystyle\sum_{k=0}^{n} \dbinom{n}{k} a^{n-k} b^k$

33. $\dbinom{m+1}{k}$

34. $\dbinom{n}{k}$

35. 0; 1; 1; 0

36. $P(A) + P(B)$

37. $1 - P(A)$

38. $P(A) + P(B) - P(A \cap B)$

39. $a_1 p_1 + a_2 p_2 + \cdots + a_n p_n$

40. $\dfrac{P(A \cap B)}{P(A)}$

41. $P(B_k|A) = \dfrac{P(A|B_k)P(B_k)}{P(A|B_1)P(B_1) + P(A|B_2)P(B_2) + \cdots + P(A|B_n)P(B_n)}$

42. $P(A \cap B) = P(A) \cdot P(B)$

43. $P(A \cap B) = P(A) \cdot P(B)$; $P(A \cap C) = P(A) \cdot P(C)$; $P(B \cap C) = P(B) \cdot P(C)$;
    $P(A \cap B \cap C) = P(A) \cdot P(B) \cdot P(C)$

Chapter 7: Functions

The aim of Section 7.1 is to promote a broad view of the function concept and to give you experience with the wide variety of functions that arise in discrete mathematics. Representation of functions by arrow diagrams is emphasized to prepare the way for the discussion of one-to-one and onto functions in Section 7.2.

Section 7.2 focuses on function properties. As you are learning about one-to-one and onto functions in this section, you may need to review the logical principles such as the negation of $\forall$, $\exists$, and if-then statements and the equivalence of a conditional statement and its contrapositive. These logical principles are needed to understand the equivalence of the two forms of the definition of one-to-one and what it means for a function not to be one-to-one or onto.

Section 7.3 on the pigeonhole principle is an application of one-to-one functions, but it is also a change of pace from the previous section of the chapter. A broad range of problems is given, from the very simple to the quite sophisticated.. It is surprising that such an "obvious" principle can be used in such a subtle way.

Sections 7.4 and 7.5 go together in the sense that the relations between one-to-one and onto functions and composition of functions developed in Section 7.4 are used to prove the fundamental theorem about cardinality in Section 7.5. The proofs that a composition of one-to-one functions is one-to-one or that a composition of onto functions is onto (and the related exercises) test the degree to which students have learned to instantiate mathematical definitions in abstract contexts, apply the method of generalizing from the generic particular in a sophisticated setting, develop mental models of mathematical concepts that are both vivid and generic enough to reason with, and create moderately complex chains of deductions.

When you read Section 7.5, try to see the connections that link Russell's paradox, the halting problem, and the Cantor diagonalization argument.

Section 7.1

3. *c.* This arrow diagram associates both 1 and 2 to 4. So this diagram does not define a function.

   d. This arrow diagram determines a function. Each element in X is related to one and only one element in Y.

   e. In this arrow diagram, the element 2 in X is not related to any element in Y. So this diagram does not define a function.

6. *b.* The answer is $2 \cdot 2 \cdot 2 \cdot 2 \cdot 2 = 2^5 = 32$. The explanation is the same as that in the answer to part (c) below, but with $n = 2$ and $m = 5$.

   c. The answer is n^m because the m elements of the domain can be placed in order and the process of constructing a function can be thought of as an m-step operation where, for each i from 1 to m, the ith step is to choose one of the n elements of the co-domain to be the image of the ith element of the domain. Since there are n ways to perform each of the m steps of the operation, the entire operation can be performed in $n \cdot n \cdots n$ (m factors) ways. (*Brief version of this explanation*: The answer is n^m because each of the m elements of the domain can be sent to any one of n possible elements in the co-domain.)

12. *b.* Define $F: \mathbf{Z}^{nonneg} \to \mathbf{R}$ as follows: for each nonnegative integer n, $F(n) = (-1)^n(2n)$.

15. *b.* $5^{-2} = 1/25$

   d. the exponent to which 3 must be raised to obtain 3^n is n

   e. $4^0 = 1$

18. *Proof*: Let b be any positive real number with $b \neq 1$. Then $b^0 = 1$, and so $\log_b 1 = 0$. (Note that we do not allow b to equal 1 here because $\log_b$ is not defined for $b = 1$.)

24. b. $mod(59,8) = 3$, $div(59,8) = 7$ c. $mod(30,5) = 0$, $div(30,5) = 6$

27. b.

```
1  2  3  4        1  2  3  4
↓  ↓  ↓  ↓        ↓  ↓  ↓  ↓
1  2  3  4        3  2  1  4
```

d.

```
1  2  3  4        1  2  3  4        1  2  3  4
↓  ↓  ↓  ↓        ↓  ↓  ↓  ↓        ↓  ↓  ↓  ↓
2  3  4  1        3  4  1  2        4  1  2  3

1  2  3  4        1  2  3  4        1  2  3  4
↓  ↓  ↓  ↓        ↓  ↓  ↓  ↓        ↓  ↓  ↓  ↓
2  1  4  3        3  1  4  2        4  3  1  2

1  2  3  4        1  2  3  4        1  2  3  4
↓  ↓  ↓  ↓        ↓  ↓  ↓  ↓        ↓  ↓  ↓  ↓
2  4  1  3        3  4  2  1        4  3  2  1
```

30. b.

input			output
x_1	x_2	x_3	f
1	1	1	1
1	1	0	1
1	0	1	0
1	0	0	0
0	1	1	1
0	1	0	1
0	0	1	0
0	0	0	0

33. f is not well defined because $f(n) \notin S$ for many values of n in S. For instance, $f(100\,000) = (100\,000)^2 = 10\,000\,000\,000 \notin S$.

39. *Proof*: Given any integer n with $n = pqr$, where p, q, and r are distinct prime numbers, let A be the set of all positive integers less than or equal to n that are divisible by p, B the set of all positive integers less than or equal to n that are divisible by q, and C the set of all positive integers less than or equal to n that are divisible by r. Note that $A = \{p, 2p, 3p, \ldots, qr \cdot p\}$, $B = \{q, 2q, 3q, \ldots, pr \cdot q\}$, $C = \{r, 2r, 3r, \ldots, pq \cdot r\}$, $A \cap B = \{pq, 2pq, 3pq, \ldots, r \cdot pq\}$, $A \cap C = \{pr, 2pr, 3pr, \ldots, q \cdot pr\}$, $B \cap C = \{qr, 2qr, 3qr, \ldots, p \cdot qr\}$, and $A \cap B \cap C = \{pqr\}$. By definition of ϕ,

$$\phi(n) = n - [n(A \cup B \cup C)]$$

$$= n - [n(A) + n(B) + n(C) - n(A \cap B) - n(A \cap C) - n(B \cap C) + n(A \cap B \cap C)]$$

by the inclusion/exclusion formula

$$= pqr - [qr + pr + pq - r - q - p + 1]$$

$$= pqr - qr - pr - pq + r + q + p - 1$$

$$= (p-1)(q-1)(r-1)$$

45. This statement is true. *Proof*: Let $f\colon X \to Y$ be any function, and suppose that $C \subseteq Y$ and $D \subseteq Y$. For any element x in X, by definition of inverse image and union,

$$x \in f^{-1}(C \cup D) \Leftrightarrow f(x) \in C \cup D \Leftrightarrow f(x) \in C \text{ or } f(x) \in D$$

$$\Leftrightarrow x \in f^{-1}(C) \text{ or } \in f^{-1}(D) \Leftrightarrow x \in f^{-1}(C) \cup f^{-1}(D).$$

Hence $f^{-1}(C \cup D) = f^{-1}(C) \cup f^{-1}(D)$.

Section 7.2

9. In each case below there are a number of correct answers.

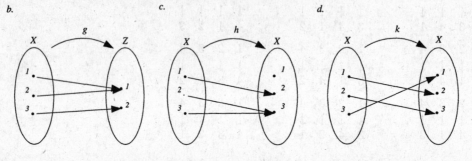

18. f is one-to-one. *Proof*: Let x_1 and x_2 be any nonzero real numbers such that $f(x_1) = f(x_2)$. By definition of f, $\dfrac{3x_1 - 1}{x_1} = \dfrac{3x_2 - 1}{x_2}$. Cross-multiplying gives $(3x_1 - 1)x_2 = (3x_2 - 1)x_1$, or, equivalently, $3x_1x_2 - x_2 = 3x_1x_2 - x_1$. Subtracting $3x_1x_2$ from both sides gives $-x_1 = -x_2$, and multiplying both sides by -1 gives $x_1 = x_2$.

24. *b. F is not onto*: The number 4 is in $\mathbf{Z}$ but $F(A) \neq 4$ for any set A in $\mathscr{P}(\{a, b, c\})$ because no subset of $\{a, b, c\}$ has four elements.

27. *a. F is one-to-one*: Suppose $F(a, b) = F(c, d)$ for some ordered pairs (a, b) and (c, d) in $\mathbf{Z}^+ \times \mathbf{Z}^+$. By definition of F, $3^a 5^b = 3^c 5^d$. Thus, by the unique factorization theorem (Theorem 3.3.3), $a = b$ and $c = d$; in other words, $(a, b) = (c, d)$.

b. G is one-to-one: Suppose $G(a, b) = G(c, d)$ for some ordered pairs (a, b) and (c, d) in $\mathbf{Z}^+ \times \mathbf{Z}^+$. By definition of G, $3^a 6^b = 3^c 6^d$, and so $3^a 3^b 2^b = 3^c 2^d 3^d$, or, equivalently, $3^{a+b} 2^b = 3^{c+d} 3^d$. Thus, by the unique factorization theorem (Theorem 3.3.3), $a + b = c + d$ and $b = d$. Solving these equations gives $a = b$ and $c = d$; in other words, $(a, b) = (c, d)$.

30. Suppose b, x, and y are any positive real numbers with $b \neq 1$. Let $u = \log_b(x)$ and $v = \log_b(y)$. By definition of logarithm, $x = b^u$ and $y = b^v$. Then $x \cdot y = b^u \cdot b^v = b^{u+v}$ by property (7.2.1). Applying the definition of logarithm to the extreme parts of this last equation gives $\log_b(x \cdot y) = u + v = \log_b(x) + \log_b(y)$.

33. When $f\colon \mathbf{R} \to \mathbf{R}$ and $g\colon \mathbf{R} \to \mathbf{R}$ are both onto, it need not be the case that $f + g$ is onto. *Counterexample*: Let $f\colon \mathbf{R} \to \mathbf{R}$ and $g\colon \mathbf{R} \to \mathbf{R}$ be defined by $f(x) = x$ and $g(x) = -x$ for all $x \in \mathbf{R}$. Then both f and g are onto, but $(f + g)(x) = f(x) + g(x) = x + (-x) = 0$ for all x, and so $f + g$ is not onto.

42. The answer to exercise 14b shows that K is onto. It is also the case that K is one-to-one. To see why this is so, suppose x_1 and x_2 are any nonnegative real numbers such that $K(x_1) = K(x_2)$. *[We must show that $x_1 = x_2$.]* Then by definition of K, $x_1^2 = x_2^2$. But each nonnegative real number has a unique nonnegative square root. So since both x_1 and x_2 are nonnegative square roots of the same number, $x_1 = x_2$ *[as was to be shown]*. Therefore K is both one-to-one and onto, and thus K is a one-to-one correspondence. For all $y \in \mathbf{R}^{nonneg}$, $K^{-1}(y) = \sqrt{y}$ because $K(\sqrt{y}) = (\sqrt{y})^2 = y$.

45. Because D is not one-to-one, D is not a one-to-one correspondence.

51. By the result of exercise 19, f is one-to-one. f is also onto for the following reason. Given any real number y other than 1, let $x = \dfrac{1+y}{1-y}$. Then x is a real number (because $y \neq 1$) and

$$f(x) = f\left(\frac{y+1}{y-1}\right) = \frac{\left(\frac{y+1}{y-1}\right)+1}{\left(\frac{y+1}{y-1}\right)-1} = \frac{\left(\frac{y+1}{y-1}\right)+1}{\left(\frac{y+1}{y-1}\right)-1}\cdot\frac{(y-1)}{(y-1)} = \frac{y+1+(y-1)}{y+1-(y-1)} = \frac{y+1+y-1}{y+1-y+1} = y.$$

This calculation also shows that $f^{-1}(y) = \dfrac{y+1}{y-1}$ for all real numbers $y \neq 1$.

54. **Algorithm 7.2.2   Checking Whether a Function is Onto**

[For a given function F with domain $X = \{a[1], a[2], \ldots, a[n]\}$ and co-domain $Y = \{b[1], b[2], \ldots, b[m]\}$, this algorithm discovers whether or not F is onto. Initially, answer is set equal to "onto", and then successive elements of Y are considered. For each such element, $b[i]$, a search is made through elements of the domain to determine if any is sent to $b[i]$. If not, the value of answer is changed to "not onto" and execution of the algorithm ceases. If so, the next successive element of Y is considered. If all elements of Y have been considered and the value of answer has not been changed from its initial value, then F is onto.]

Input: *n [a positive integer], $a[1], a[2], \ldots, a[n]$ [a one-dimensional array representing the set X], m [a positive integer], $b[1], b[2], \ldots, b[m]$ [a one-dimensional array representing the set Y], F [a function with domain X]*

Algorithm Body:

```
answer := "onto"
i := 1
while (i ≤ m and answer = "onto")
       j := 1
       found := "no"
       while (j ≤ n and found = "no")
              if F(a[j]) = b[i] then found := "yes"
              j := j + 1
       end while
       if found = "no" then answer := "not onto"
       i := i + 1
end while
```

Output: *answer [a string]*

Section 7.3

6. *a.* Yes. Let X be the set of seven integers and Y the set of all possible remainders obtained through division by 6, and consider the function R from X (the pigeons) to Y (the pigeonholes) defined by the rule: $R(n) = n \bmod 6$ (= the remainder obtained by the integer division of n by 6). Now X has 7 elements and Y has 6 elements (0, 1, 2, 3, 4, and 5). Hence by the pigeonhole principle, R is not one-to-one: $R(n_1) = R(n_2)$ for some integers n_1 and n_2 with $n_1 \neq n_2$. But this means that n_1 and n_2 have the same remainder when divided by 6.

b. No. Consider the set $\{1, 2, 3, 4, 5, 6, 7\}$. This set has seven elements no two of which have the same remainder when divided by 8.

15. There are $n+1$ even integers from 0 to $2n$ inclusive: $0\,(=2\cdot0), 2\,(=2\cdot1), 4\,(=2\cdot2), \ldots, 2n\,(=2\cdot n)$. So a maximum of $n+1$ even integers can be chosen. Thus if at least $n+2$ integers are chosen, one is sure to be odd. Similarly, there are n odd integers from 0 to $2n$ inclusive, namely $1\,(=2\cdot1-1), 3\,(=2\cdot2-1), \ldots, 2n-1\,(=2\cdot n-1)$. It follows that if at least $n+1$ integers are chosen, one is sure to be even. (An alternative way to reach the second conclusion is to note that there are $2n+1$ integers from 0 to $2n$ inclusive. Because $n+1$ of them are even, the number of odd integers is $(2n+1)-(n+1)=n$.)

18. There are 15 distinct remainders that can be obtained through integer division by 15 (0, 1, 2, $\ldots$, 14). Hence at least 16 integers must be chosen in order to be sure that at least two have the same remainder when divided by 15.

21. The length of the repeating section of the decimal representation of 5/20483 is less than or equal to the number of possible remainders that can be obtained when a number is divided by 20,483, , namely 20,483. The reason is that in the long-division process of $5.0000\ldots$ by 20,483, either some remainder is 0 and the decimal expansion terminates (in which case the length of the repeating section is 0) or, at some point within the first 20,483 successive divisions in the long-division process, a nonzero remainder is repeated. At that point the digits in the developing decimal expansion begin to repeat because the sequence of successive remainders repeats those previously obtained.

27. Yes. Let X be the set of 2,000 people (the pigeons) and Y the set of all 366 possible birthdays (the pigeonholes). Define a function $B: X \to Y$ by specifying that $B(x) = x$'s birthday. Now $2000 > 4\cdot366 = 1464$, and so by the generalized pigeonhole principle, there must be some birthday y such that $B^{-1}(y)$ has at least $4+1=5$ elements. Hence at least 5 people must share the same birthday.

30. Consider the maximum number of pennies that can be chosen without getting at least five from the same year. This maximum, which is 12, is obtained when four pennies are chosen from each of the three years. Hence at least thirteen pennies must be chosen to be sure of getting at least five from the same year.

33. *Proof*: Let T be the set of all possible sums of elements of subsets of S and define a function F from $\mathscr{P}(S)$ to T as follows: for each subset X of A, let $F(X)$ be the sum of the elements of X. By Theorem 5.3.1, $\mathscr{P}(S)$ has $2^{10} = 1024$ elements. Moreover, because S has 10 elements each of which is less than 50, the maximum possible sum of elements of any subset of S is $41 + 42 + \cdots + 50 = 455$. The minimum possible sum of elements of any subset of S is $1 + 2 + \cdots + 10 = 55$. Hence T has $455 - 55 + 1 = 401$ elements (the numbers from 55 to 455 inclusive). Because $1024 > 401$, the pigeonhole principle guarantees that F is not one-to-one. Thus there exist distinct subsets S_1 and S_2 of S such that $F(S_1) = F(S_2)$, which implies that the elements of S_1 add up to the same sum as the elements of S_2.

    Additional Note: In fact, it can be shown that it is always possible to find disjoint subsets of S with the same sum. To see why this is true, consider again the sets S_1 and S_2 found in the proof given above. Then $S_1 \neq S_2$ and $F(S_1) = F(S_2)$. By definition of F, $F(S_1 - S_2) + F(S_1 \cap S_2) =$ the sum of the elements in $S_1 - S_2$ plus the sum of the elements in $S_1 \cap S_2$. But $S_1 - S_2$ and $S_1 \cap S_2$ are disjoint and their union is S_1. So $F(S_1 - S_2) + F(S_1 \cap S_2) = F(S_1)$. By the same reasoning, $F(S_2 - S_1) + F(S_1 \cap S_2) = F(S_2)$. Since $F(S_1) = F(S_2)$, we have that $F(S_1 - S_2) = F(S_1) - F(S_1 \cap S_2) = F(S_2) - F(S_1 \cap S_2) = F(S_2 - S_1)$. Hence the elements in $S_1 - S_2$ add up to the same sum as the elements in $S_2 - S_1$. But $S_1 - S_2$ and $S_2 - S_1$ are disjoint because $S_1 - S_2$ contains no elements of S_2 and $S_2 - S_1$ contains no elements of S_1.

36. a. *Proof*: Suppose $a_1, a_2, \ldots, a_n$ is a sequence of n integers none of which is divisible by n. Define a function F from $X = \{a_1, a_2, \ldots, a_n\}$ to $Y = \{1, 2, \ldots, n-1\}$ by the rule $F(x) = x \bmod n$ (the remainder obtained through integer division of x by n). Since no element of X is divisible by n, F is well-defined. Now X has n elements and Y has $n-1$ elements, and so

by the pigeonhole principle $F(a_i) = F(a_j)$ for some elements a_i and a_j in X with $a_i \neq a_j$. By definition of F, both a_i and a_j have the same remainder when divided by n, and so $a_i - a_j$ is divisible by n. [More formally, by the quotient-remainder theorem we can write $a_i = nq_i + r_i$ and $a_j = nq_j + r_j$ where $0 \leq r_i < n$ and $0 \leq r_j < n$, and since $F(a_i) = F(a_j)$, $r_i = r_j$. Thus $a_i - a_j = (nq_i + r_i) - (nq_j + r_j) = n(q_i - q_j) + (r_i - r_j) = n(q_i - q_j)$ because $r_i = r_j$. So by definition of divisibility, $a_i - a_j$ is divisible by n.]

b. *Proof:* Suppose $x_1, x_2, \ldots, x_n$ is a sequence of n integers. For each $k = 1, 2, \ldots, n$, let $a_k = x_1 + x_2 + \ldots + x_k$. If some a_k is divisible by n, the problem is solved: the sum of the numbers in the consecutive subsequence $x_1, x_2, \ldots, x_k$ is divisible by n. If no a_k is divisible by n, then $a_1, a_2, \ldots, a_n$ satisfies the hypothesis of part a, and so $a_j - a_i$ is divisible by n for some integers i and j with $j > i$. But $a_j - a_i = x_{i+1} + x_{i+2} + \ldots + x_j$. Thus the sum of the numbers in the consecutive subsequence $x_{i+1}, x_{i+2}, \ldots, x_j$ is divisible by n.

39. **Algorithm 7.3.1   Finding Pigeons in the Same Pigeonhole**

[For a given function F with domain $X = \{x[1], x[2], \ldots, x[n]\}$ and co-domain $Y = \{y[1], y[2], \ldots, y[m]\}$ with $n > m$, this algorithm finds elements a and b so that $F(a) = F(b)$ and $a \neq b$. The existence of such elements is guaranteed by the pigeonhole principle because $n > m$. Initially, the variable done is set equal to "no". Then the values of $F(x[i])$ and $F(x[j])$ are systematically compared for indices i and j with $1 \leq i < j \leq n$. When it is found that $F(x[i]) = F(x[j])$ and $x[i] \neq x[j]$, a is set equal to $x[i]$, b is set equal to $x[j]$, done is set equal to "yes", and execution ceases.]

Input: n [a positive integer], m [a positive integer with $m < n$], $x[1], x[2], \ldots, x[n]$ [a one-dimensional array representing the set X], $y[1], y[2], \ldots, y[m]$ [a one-dimensional array representing the set Y], F [a function from X to Y]

Algorithm Body:

$done := \text{"no"}$

$i := 1$

while $(i \leq n - 1 \text{ and } done = \text{"no"})$

    $j := i + 1$

    while $(j \leq n \text{ and } done = \text{"no"})$

        if $(F(x[i]) = F(x[j]) \text{ and } x[i] \neq x[j])$

            then do $a := x[i]$, $b := x[j]$, $done := \text{"yes"}$ **end do**

        $j := j + 1$

    end while

    $i := i + 1$

end while

Output: a, b [positive integers]

Section 7.4

6. $G \circ F$ is defined by $(G \circ F)(x) = \lfloor x \rfloor$, for all real numbers x, because for any real number x, $(G \circ F)(x) = G(F(x)) = G(3x) = \left\lfloor \dfrac{3x}{3} \right\rfloor = \lfloor x \rfloor$. $F \circ G$ is defined by $(F \circ G)(n) = 3 \cdot \left\lfloor \dfrac{x}{3} \right\rfloor$, for all real numbers x, because for any real number x, $(F \circ G)(x) = F(G(x)) = F(\left\lfloor \dfrac{x}{3} \right\rfloor) = 3 \cdot \left\lfloor \dfrac{x}{3} \right\rfloor$.

Then $G \circ F \neq F \circ G$ because, for instance, $(G \circ F)(1) = \lfloor 1 \rfloor = 1$, whereas $(F \circ G)(1) = 3 \cdot \left\lfloor \dfrac{1}{3} \right\rfloor = 3 \cdot 0 = 0$.

12. *b.* For all positive real numbers b and x, $\log_b x$ is the exponent to which b must be raised to obtain x. So if b is raised to this exponent, x is obtained. In other words, $b^{\log_b x} = x$.

15. *b.* $z/2 = t/2$ *c.* $f(x_1) = f(x_2)$

18. Yes. *Proof:* Suppose $f: X \to Y$ and $g: Y \to Z$ are functions and $g \circ f: X \to Z$ is one-to-one. To show that f is one-to-one, suppose x_1 and x_2 are in X and $f(x_1) = f(x_2)$. *[We must show that $x_1 = x_2$.]* Then $g(f(x_1)) = g(f(x_2))$, and so $g \circ f(x_1) = g \circ f(x_2)$. But $g \circ f$ is one-to-one. Hence $x_1 = x_2$ *[as was to be shown]*.

24. $g \circ f: \mathbf{R} \to \mathbf{R}$ is defined by $(g \circ f)(x) = g(f(x)) = g(x+3) = -(x+3)$ for all $x \in \mathbf{R}$.

Since $z = -(x+3)$ if, and only if, $x = -z - 3$, $(g \circ f)^{-1}: \mathbf{R} \to \mathbf{R}$ is defined by $(g \circ f)^{-1}(z) = -z - 3$ for all $z \in \mathbf{R}$.

Since $z = -y$ if, and only if, $y = -z$, $g^{-1}: \mathbf{R} \to \mathbf{R}$ is defined by $g^{-1}(z) = -z$ for all $z \in \mathbf{R}$.

Since $y = x + 3$ if, and only if, $x = y - 3$, $f^{-1}: \mathbf{R} \to \mathbf{R}$ is defined by $f^{-1}(y) = y - 3$.

$f^{-1} \circ g^{-1}: \mathbf{R} \to \mathbf{R}$ is defined by $(f^{-1} \circ g^{-1})(z) = f^{-1}(g^{-1}(z)) = f^{-1}(-z) = (-z) - 3 = -z - 3$ for all $z \in \mathbf{R}$.

By the above and the definition of equality of functions, $(g \circ f)^{-1} = f^{-1} \circ g^{-1}$.

30. False. *One counterexample among many:* Let $X = Y = C = \{1, 2\}$, and define $f : X \to Y$ by specifying that $f(1) = f(2) = 1$. Then $f(f^{-1}(C)) = f(\{1, 2\}) = \{1\}$. So $C \not\subseteq f(f^{-1}(C))$ because $\{1, 2\} \not\subseteq \{1\}$.

Section 7.5

6. The function $I: 2\mathbf{Z} \to \mathbf{Z}$ is defined as follows: $I(n) = n$ for all even integers n. I is clearly one-to-one because if $I(n_1) = I(n_2)$ then by definition of I, $n_1 = n_2$. But I is not onto because the range of I consists only of even integers. In other words, if m is any odd integer, then $I(n) \neq m$ for any even integer n.

The function $J: \mathbf{Z} \to 2\mathbf{Z}$ is defined as follows $J(n) = 2 \lfloor n/2 \rfloor$ for all integers n. Then J is onto because for any even integer m, $m = 2k$ for some integer k. Let $n = 2k$. Then $J(n) = J(2k) = 2 \lfloor 2k/2 \rfloor = 2 \lfloor k \rfloor = 2k = m$. But J is not one-to-one because, for example, $J(2) = 2 \lfloor 2/2 \rfloor = 2 \cdot 1 = 2$ and $J(3) = 2 \lfloor 3/2 \rfloor = 2 \cdot 1 = 2$, so $J(2) = J(3)$ but $2 \neq 3$.

(More generally, given any integer k, if $m = 2k$, then $J(m) = 2 \lfloor m/2 \rfloor = 2 \lfloor 2k/2 \rfloor = 2 \lfloor k \rfloor = J(m)$ and $J(m+1) = 2 \lfloor (m+1)/2 \rfloor = 2 \lfloor (2k+1)/2 \rfloor = 2 \lfloor k + 1/2 \rfloor = 2k$. So $J(m) = J(m+1)$ but $m \neq m + 1$.)

9. Define a function $f: \mathbf{Z}^+ \to \mathbf{Z}^{nonneg}$ as follows: $f(n) = n - 1$ for all positive integers n. Observe that if $n \geq 1$ then $n - 1 \geq 0$, so f is well-defined. In addition, f is one-to-one because for all positive integers n_1 and n_2, if $f(n_1) = f(n_2)$ then $n_1 - 1 = n_2 - 1$ and hence $n_1 = n_2$. Moreover f is onto because if m is any nonnegative integer, then $m + 1$ is a positive integer and $f(m+1) = (m+1) - 1 = m$ by definition of f. Thus, because there is a function $f: \mathbf{Z}^+ \to \mathbf{Z}^{nonneg}$ that is one-to-one and onto, $\mathbf{Z}^+$ has the same cardinality as $\mathbf{Z}^{nonneg}$. It follows that $\mathbf{Z}^{nonneg}$ is countably infinite and hence countable.

12. *Proof:* Define $F: S \to W$ by the rule $F(x) = (b-a)x + a$ for all real numbers x in S. Then F is well-defined because if $0 < x < 1$, then $a < (b-a)x + a < b$. In addition, F is one-to-one because if x_1 and x_2 are in S and $F(x_1) = F(x_2)$, then $(b-a)x_1 + a = (b-a)x_2 + a$ and so *[by subtracting a and dividing by $b-a$]* $x_1 = x_2$. Furthermore, F is onto because if y is any element in W, then $a < y < b$ and so $0 < (y-a)/(b-a) < 1$. Consequently, $(y-a)/(b-a) \in S$ and $h((y-a)/(b-a)) = (b-a)[(y-a)/(b-a)] + a = y$. Hence F is a one-to-one correspondence, and so S and W have the same cardinality.

15. Let B be the set of all bit strings (strings of 0's and 1's). Define a function $F: \mathbf{Z}^+ \to B$ as follows: $F(1) = \epsilon$, $F(2) = 0$, $F(3) = 1$, $F(4) = 00$, $F(5) = 01$, $F(6) = 10$, $F(7) = 11$, $F(8) = 000$, $F(9) = 001$, $F(10) = 010$, and so forth. At each stage, all the strings of length k are counted before the strings of length $k+1$, and the strings of length k are counted in order of increasing magnitude when interpreted as binary representations of integers. Thus the set of all bit strings is countably infinite and hence countable.

Note: A more formal definition for F is the following:

$$F(n) = \begin{cases} \epsilon & \text{if } n = 1 \\ \text{the } k\text{-bit binary representation of } n - 2^k & \text{if } \lfloor \log_2 n \rfloor = k. \end{cases}$$

For example, $F(7) = 11$ because $\lfloor \log_2 7 \rfloor = 2$ and the two-bit binary representation of $7 - 2^2$ $(= 3)$ is 11.

18. No. For instance, both $\sqrt{2}$ and $-\sqrt{2}$ are irrational (by Theorem 3.7.1 and exercise 21 in Section 3.6), and yet their average is $(\sqrt{2} + (-\sqrt{2}))/2$ which equals 0 and is rational.

More generally: If r is any rational number and x is any irrational number, then both $r+x$ and $r-x$ are irrational (by the result of exercise 11 in Section 3.6 or by the combination of Theorem 3.6.3 and exercise 9 in Section 3.6). Yet the average of these numbers is $((r+x)+(r-x))/2 = r$, which is rational.

21. *Two examples of many*: Define $F: \mathbf{Z} \to \mathbf{Z}$ by the rule $F(n) = \begin{cases} n/2 & \text{if } n \text{ is even} \\ 0 & \text{if } n \text{ is odd} \end{cases}$. Then F is onto because given any integer m, $m = F(2m)$. But F is not one-to-one because, for instance, $F(1) = F(3) = 0$.

Define $G: \mathbf{Z} \to \mathbf{Z}$ by the rule $G(n) = \lfloor n/2 \rfloor$ for all integers n. Then G is onto because given any integer m, $m = \lfloor m \rfloor = \lfloor (2m)/2 \rfloor = G(2m)$. But G is not one-to-one because, for instance, $G(2) = \lfloor 2/2 \rfloor = 1$ and $G(3) = \lfloor 3/2 \rfloor = 1$ and $2 \neq 3$.

24. The proof given below is adapted from one in *Foundations of Modern Analysis* by Jean Dieudonné, New York: Academic Press, 1969, page 14.

Proof: Suppose (a, b) and (c, d) are in $\mathbf{Z}^+ \times \mathbf{Z}^+$ and $(a, b) \neq (c, d)$.

Case 1, $a + b \neq c + d$: By interchanging (a, b) and (c, d) if necessary, we may assume that $a + b < c + d$. Then

$$
\begin{aligned}
H(a, b) &= b + \frac{(a+b)(a+b+1)}{2} && \text{by definition of } H \\
\Rightarrow \quad H(a, b) &\leq a + b + \frac{(a+b)(a+b+1)}{2} && \text{because } a \geq 0 \\
\Rightarrow \quad H(a, b) &< (a+b+1) + \frac{(a+b)(a+b+1)}{2} && \text{because } a + b < a + b + 1 \\
\Rightarrow \quad H(a, b) &< \frac{2(a+b+1)}{2} + \frac{(a+b)(a+b+1)}{2} && \\
\Rightarrow \quad H(a, b) &< \frac{(a+b+1)(a+b+2)}{2} && \text{by factoring out } (a+b+1) \\
\Rightarrow \quad H(a, b) &< \frac{(c+d)(c+d+1)}{2} && \text{since } a+b < c+d \text{ and } a, b, c, \text{ and} \\
& && d \text{ are integers, } a+b+1 \leq c+d \\
\Rightarrow \quad H(a, b) &< d + \frac{(c+d)(c+d+1)}{2} && \text{because } d \geq 0 \\
\Rightarrow \quad H(a, b) &< H(c, d) && \text{by definition of } H.
\end{aligned}
$$

Therefore, $H(a, b) \neq H(c, d)$.

Case 2, $a + b = c + d$: First observe that in this case $b \neq d$. For if $b = d$, then subtracting b from both sides of $a + b = c + d$ gives $a = c$, and so $(a, b) = (c, d)$, which contradicts our assumption that $(a, b) \neq (c, d)$. Hence,

$$H(a, b) = b + \frac{(a + b)(a + b + 1)}{2} = b + \frac{(c + d)(c + d + 1)}{2} \neq d + \frac{(c + d)(c + d + 1)}{2} = H(c, d),$$

and so $H(a, b) \neq H(c, d)$.

Thus both in case 1 and in case 2, $H(a, b) \neq H(c, d)$, and hence H is one-to-one.

27. *Proof*: Suppose A is any countable set, B is any set, and $g \colon A \to B$ is onto. Since A is countable, there is a one-to-one correspondence $f \colon \mathbf{Z}^+ \to A$. Then, in particular, f is onto, and so by Theorem 7.4.4, $g \circ f$ is an onto function from $\mathbf{Z}^+$ to B. Define a function $h \colon B \to \mathbf{Z}^+$ as follows: Suppose x is any element of B. Since $g \circ f$ is onto, $\{ m \in \mathbf{Z}^+ \mid (g \circ f)(m) = x \} \neq \emptyset$. So by the well-ordering principle for the integers, this set has a least element. In other words, there is a least positive integer n with $(g \circ f)(m) = x$. Let $h(x)$ be this integer.

We claim that h is one-to-one. For suppose $h(x_1) = h(x_2) = n$. By definition of h, n is the least positive integer with $(g \circ f)(m) = x_1$. But also by definition of h, n is the least positive integer with $(g \circ f)(m) = x_2$. So $x_1 = (g \circ f)(m) = x_2$.

Thus h is a one-to-one correspondence between B and a subset S of positive integers (the range of h). Since any subset of a countable set is countable (Theorem 7.5.3), S is countable, and so there is a one-to-one correspondence between B and a countable set. Hence by the transitive property of cardinality B is countable.

30. *Proof*: Suppose A is any finite set and B is any countably infinite set and A and B are disjoint. In case $A = \emptyset$, then $A \cup B = B$, which is countably infinite. So we may assume that for some positive integer m there are one-to-one correspondences $f \colon \{1, 2, \ldots, m\} \to A$ and $g \colon \mathbf{Z}^+ \to B$. Define a function $h \colon \mathbf{Z}^+ \to A \cup B$ as follows:

$$\text{For all integers } n, \ h(n) = \begin{cases} f(n) & \text{if } 1 \leq n \leq m \\ g(n - m) & \text{if } n \geq m + 1 \end{cases}.$$

Then h is one-to-one because $A \cap B = \emptyset$ and f and g are one-to-one. And h is onto because both f and g are onto and every positive integer can be written in the form $n - m$ for some integer $n \geq m + 1$. Since h is one-to-one and onto, h is a one-to-one correspondence. Therefore, $A \cup B$ is countably infinite.

33. *Proof*: First note that there are as many equations of the form $x^2 + bx + c = 0$ as there are pairs (b, c) where b and c are in $\mathbf{Z}$. By exercise 32, the set of all such pairs is countably infinite, and so the set of equations of the form $x^2 + bx + c = 0$ is countably infinite.

Next observe that, by the quadratic formula, each equation $x^2 + bx + c = 0$ has at most two solutions (which may be complex numbers):

$$x = \frac{-b + \sqrt{b^2 - 4c}}{2} \quad \text{and} \quad x = \frac{-b - \sqrt{b^2 - 4c}}{2}.$$

Let

$$R_1 = \left\{ x \mid x = \frac{-b + \sqrt{b^2 - 4c}}{2} \quad \text{for some integers } b \text{ and } c \right\},$$

$$R_2 = \left\{ x \mid x = \frac{-b - \sqrt{b^2 - 4c}}{2} \quad \text{for some integers } b \text{ and } c \right\},$$

and $R = R_1 \cup R_2$. Then R is the set of all solutions of equations of the form $x^2 + bx + c = 0$ where b and c are integers.

Define functions F_1 and F_2 from the set of equations of the form $x^2 + bx + c = 0$ to the sets R_1 and R_2 as follows:

$$F_1(x^2 + bx + c = 0) = \frac{-b + \sqrt{b^2 - 4c}}{2} \quad \text{and} \quad F_2(x^2 + bx + c = 0) = \frac{-b - \sqrt{b^2 - 4c}}{2}.$$

Then F_1 and F_2 are onto functions defined on countably infinite sets, and so, by exercise 27, R_1 and R_2 are countable. Since any union of two countable sets is countable (exercise 31), $R = R_1 \cup R_2$ is countable.

36. *Proof*: Let B be the set of all functions from $\mathbf{Z}^+$ to $\{0, 1\}$ and let D be the set of all functions from $\mathbf{Z}^+$ to $\{0, 1, 2, 3, 4, 5, 6, 7, 8, 9\}$. Elements of B can be represented as infinite sequences of 0's and 1's (for instance, 01101010110...) and elements of D can be represented as infinite sequences of digits from 0 to 9 inclusive (for instance, 20775931124...).

We define a function $H: B \to D$ as follows: For each function f in B, consider the representation of f as an infinite sequence of 0's and 1's. Such a sequence is also an infinite sequence of digits chosen from 0 to 9 inclusive (one formed without using 2,3,...,9), which represents a function in D. We define this function to be $H(f)$. More formally, for each $f \in B$, let $H(f)$ be the function in D defined by the rule $H(f)(n) = f(n)$ for all $n \in \mathbf{Z}^+$. It is clear from the definition that H is one-to-one.

We define a function $K: D \to B$ as follows: For each function g in D, consider the representation of g as a sequence of digits from 0 to 9 inclusive. Replace each of these digits by its 4-bit binary representation adding leading 0's if necessary to make a full four bits. (For instance, 2 would be replaced by 0010.) The result is an infinite sequence of 0's and 1's, which represents a function in B. This function is defined to be $K(g)$. Note that K is one-to-one because if $g_1 \neq g_2$ then the sequences representing g_1 and g_2 must have different digits in some position m, and so the corresponding sequences of 0's and 1's will differ in at least one of the positions $4m - 3, 4m - 2, 4m - 1$, or $4m$, which are the locations of the 4-bit binary representations of the digits in position m.

It can be shown that whenever there are one-to-one functions from one set to a second and from the second set back to the first, then the two sets have the same cardinality. This fact is known as the Schröder-Bernstein theorem after its two discoverers. For a proof see, for example, *Set Theory and Metric Spaces* by Irving Kaplansky, *A Survey of Modern Algebra*, Third Edition, by Garrett Birkhoff and Saunders MacLane, *Naive Set Theory* by Paul Halmos, or *Topology* by James R. Munkres. The above discussion shows that there are one-to-one functions from B to D and from D to B, and hence by the Schröder-Bernstein theorem the two sets have the same cardinality.

General Review for Chapter 7

Definitions: How are the following terms defined?

- function f from a set X to a set Y *(p. 390)*
- If f is a function from a set X to a set Y, what are

  - the domain, co-domain, and range of f *(p. 390)*

  - the image of X under f *(p. 390)*

  - the value of f at x, where x is in X *(p. 390)*

  - the image of x under f, where x is in X *(p. 390)*

  - the output of f for the input x, where x is in X *(p. 390)*

  - an inverse image of y, where y is in Y *(p. 390)*

- logarithm with base b of a positive number x *(p. 395)*
- one-to-one function *(p. 402)*
- onto function *(p. 407)*
- exponential function with base b *(p. 411)*
- one-to-one correspondence *(p. 413)*
- inverse function *(p. 415)*
- composition of functions *(p. 432)*
- cardinality *(p. 443)*
- countable set and uncountable set. *(p. 445)*

General Function Facts

- How do you draw an arrow diagram for a function defined on a finite set? *(p. 390)*
- Given a function defined by an arrow diagram or by a formula, how do you find values of the function, the range of the function, and the inverse image of an element in its co-domain? *(p. 391, pp. 394-8)*
- How do you show that two functions are equal? *(p. 393)*
- If the claim is made that a given formula defines a function from a set X to a set Y, how do you determine that the "function" is not well-defined? *(p. 398)*

One-to-one and Onto

- How do you show that a function is not one-to-one? *(p. 403, 404)*
- How do you show that a function defined on an infinite set is one-to-one? *(p. 404)*
- How do you show that a function is not onto? *(p. 407, 409)*
- How do you show that a function defined on an infinite set is onto? *(p. 409)*
- How do you determine if a given function has an inverse function? *(p. 415)*
- How do you find an inverse function if it exists? *(p. 415-6)*

Exponents and Logarithms

- What are the four laws of exponents? *(p. 411)*
- What are the corresponding properties of logarithms? *(p. 412 and 419-exercises 29-31)*

- How are the logarithmic function with base b and the exponential function with base b related? *(p. 415)*

Composition of Functions

- How do you compute the composition of two functions? *(p. 432)*
- What kind of function do you obtain when you compose two one-to-one functions? *(p. 437)*
- What kind of function do you obtain when you compose two onto functions? *(p. 438)*
- What kind of function do you obtain when you compose a one-to-one function with a function that is not one-to-one? *(p. 442-exercise 18)*
- What kind of function do you obtain when you compose an onto function with a function that is not onto? *(p. 442-exercise 19)*
- What is the composition of a function with its inverse? *(p. 436)*

Applications of Functions

- What is the pigeonhole principle? *(p. 420)*
- What is the generalized pigeonhole principle? *(p. 425)*
- How do you show that one set has the same cardinality as another? *(p. 443)*
- How do you show that a given set is countably infinite? countable? *(p. 446)*
- How do you show that the set of all positive rational numbers is countable? *(p. 448)*
- How is the Cantor diagonalization process used to show that the set of real numbers between 0 and 1 is uncountable? *(p. 450)*

Test Your Understanding: Chapter 7

Test yourself by filling in the blanks.

1. A function f from a set X to a set Y is a relation between elements of X (called inputs) and elements of Y (called outputs) such that _____ input element of X is related to _____ output element of Y.

2. Given a function f from a set X to a set Y, $f(x)$ is _____.

3. Given a function f from a set X to a set Y, if $f(x) = y$, then y is called _____ or _____ or _____ or _____.

4. Given a function f from a set X to a set Y, the range of f (or the image of X under f) is _____.

5. Given a function f from a set X to a set Y, if $f(x) = y$, then x is called _____ or _____.

6. Given a function f from a set X to a set Y, if $y \in Y$, then $f^{-1}(y) =$ _____ and is called _____.

7. Given functions f and g from a set X to a set Y, $f = g$ if, and only if, _____.

8. Given positive real numbers x and b with $b \neq 1$, $\log_b x =$ _____.

9. If F is a function from a set X to a set Y, then F is one-to-one if, and only if, _____.

10. If F is a function from a set X to a set Y, then F is not one-to-one if, and only if, _____.

11. If F is a function from a set X to a set Y, then F is onto if, and only if, _____.

12. If F is a function from a set X to a set Y, then F is not onto if, and only if, ____.

13. The following two statements are ____:

    $\forall\ u,v \in U$, if $H(u) = H(v)$ then $u = v$.

    $\forall\ u,v \in U$, if $u \neq v$ then $H(u) \neq H(v)$.

14. Given a function $F\colon X \to Y$ (where X is an infinite set or a large finite set), to prove that F is one-to-one, you suppose that ____ and then you show that ____.

15. Given a function $F\colon X \to Y$) (where X is an infinite set or a large finite set), to prove that F is onto, you suppose that ____ and then you show that ____.

16. Given a function $F\colon X \to Y$, to prove that F is not one-to-one, you ____.

17. Given a function $F\colon X \to Y$, to prove that F is not onto, you ____.

18. A one-to-one correspondence from a set X to a set Y is a ____ that is ____.

19. If F is a one-to-one correspondence from a set X to a set Y and y is in Y, then $F^{-1}(y)$ is ____.

20. The pigeonhole principle states that ____.

21. The generalized pigeonhole principle states that ____.

22. If X and Y are finite sets and f is a function from X to Y then f is one-to-one if, and only if, ____.

23. If f is a function from X to Y and g is a function from Y to Z, then $g \circ f$ is a function from ____ to ____, and $(g \circ f)(x)$ ____ for all x in X.

24. If f is a function from X to Y and i_X and i_Y are the identity functions from X to X and Y to Y, respectively, then $f \circ i_X =$ ____ and $i_Y \circ f =$ ____.

25. If f is a one-to-one correspondence from X to Y, then $f^{-1} \circ f =$ ____ and $f \circ f^{-1} =$ ____.

26. If f is a one-to-one function from X to Y and g is a one-to-one function from Y to Z, you prove that $g \circ f$ is one-to-one by supposing that ____ and then showing that ____.

27. If f is an onto function from X to Y and g is an onto function from Y to Z, you prove that $g \circ f$ is onto by supposing that ____ and then showing that ____.

28. A set is finite if, and only if, ____.

29. To prove that a set A has the same cardinality as a set B you must ____.

30. Given a set A, the reflexive property of cardinality says that ____.

31. Given sets A and B, the symmetric property of cardinality says that ____.

32. Given sets A, B, and C, the transitive property of cardinality says that ____.

33. A set is called countably infinite if, and only if, ____.

34. A set is called countable if, and only if, ____.

35. In each of the following, fill in the blank with the word countable or the word uncountable.

    (a) The set of all integers is ____.

    (b) The set of all rational numbers is ____.

(c) The set of all real numbers between 0 and 1 is ____.

(d) The set of all real numbers is ____.

(e) The set of all computer programs in a given computer language is ____.

(f) The set of all functions from the set of all positive integers, $\mathbf{Z}^+$, to $\{0, 1, 2, 3, 4, 5, 6, 7, 8, 9\}$ is ____.

36. The Cantor diagonalization process is used to prove that ____.

Answers

1. each, one and only one

2. the unique output element y in Y that is related to x by f

3. the value of f at x; the image of x under f; the output of f for the input x

4. the set of all y in Y such that $f(x) = y$

5. an inverse image of y under f; a preimage of y

6. $\{x \in X \mid f(x) = y\}$; the inverse image of y

7. $f(x) = g(x)$ for all $x \in X$

8. the exponent to which b must be raised to obtain x
   Or: $\log_b y = x \Leftrightarrow b^y = x$

9. for all x_1 and x_2 in X, if $F(x_1) = F(x_2)$ then $x_1 = x_2$

10. there exist elements x_1 and x_2 in X such that $F(x_1) = F(x_2)$ and $x_1 \neq x_2$

11. for all y in Y, there exists at least one element x in X such that $f(x) = y$

12. there exists an element y in Y such that for all elements x in X, $f(x) \neq y$

13. logically equivalent ways of expression what it means for H to be a one-to-one function (The second way is the contrapositive of the first.)

14. x_1 and x_2 are any *[particular but arbitrarily chosen]* elements in X with the property that $F(x_1) = F(x_2)$; $x_1 = x_2$

15. y is any *[particular but arbitrarily chosen]* element in Y; there exists at least one element x in X such that $F(x) = y$

16. show that there are concrete elements x_1 and x_2 in X with the property that $F(x_1) = F(x_2)$ and $x_1 \neq x_2$

17. show that there is a concrete element y in Y with the property that $F(x) \neq y$ for any element x in X

18. function from X to Y; one-to-one and onto

19. the unique element x in X such that $F(x) = y$ (in other words, $F^{-1}(y)$ is the unique preimage of y in X)

20. if n pigeons fly into m pigeonholes and $n > m$, then at least two pigeons fly into the same pigeonhole
    Or: given any function from a finite set to a smaller finite set, there must be at least two elements in the function's domain that have the same image in the function's co-domain
    Or: a function from one finite set to a smaller finite set cannot be one-to-one

21. if n pigeons fly into m pigeonholes and, for some positive integer k, $n > mk$, the at least one pigeonhole contains $k + 1$ or more pigeons
    Or: for any function f from a finite set X to a finite set Y and for any positive integer k, if $N(X) > k \cdot N(Y)$, then there is some $y \in Y$ such that y is the image of at least $k + 1$ distinct elements of Y

22. f is onto

23. X; Z; $g(f(x))$

24. f; f

25. i_X; i_Y

26. x_1 and x_2 are any *[particular but arbitrarily chosen]* elements in X with the property that $(g \circ f)(x_1) = (g \circ f)(x_2)$; $x_1 = x_2$

27. z is any *[particular but arbitrarily chosen]* element in Z; there exists at least one element x in X such that $(g \circ f)(x) = z$

28. it is the empty set or there is a one-to-one correspondence from $\{1, 2, \ldots, n\}$ to it, where n is a positive integer

29. show that there exists a function from A to B that is one-to-one and onto;
    Or: show that there exists a one-to-one correspondence from A to B

30. A has the same cardinality as A

31. if A has the same cardinality as B, then B has the same cardinality as A

32. if A has the same cardinality as B and B has the same cardinality as C, then A has the same cardinality as C

33. it has the same cardinality as the set of all positive integers

34. it is finite or countably infinite

35. countable; countable; uncountable; uncountable; countable; uncountable

36. the set of all real numbers between 0 and 1 is uncountable

Chapter 8: Recursion

Section 8.1 has two aims. The primary one is to introduce the idea of "recursive thinking," namely assuming the answer to a problem is known for certain smaller cases and expressing the answer for a given case in terms of the answers to these smaller cases. The other related aim is simply to help you become familiar with the concept and notation of a recursively defined sequence.

Sections 8.2 and 8.3 treat the questions of how to find an explicit formula for a sequence that has been defined recursively and how to use mathematical induction to verify that this explicit formula correctly describes the given sequence. In Section 8.2, the method used is "iteration," which consists of writing down successive terms of the sequence and looking for a pattern. In Section 8.3, explicit formulas for second-order linear homogeneous recurrence relations are derived.

Section 8.4 introduces the very important concepts of general recursive functions, general recursive definitions for the elements of a set, and the related notion of structural induction.

Section 8.1

6. $t_0 = -1$, $t_1 = 2$, $t_2 = t_1 + 2 \cdot t_0 = 2 + 2 \cdot (-1) = 0$, $t_3 = t_2 + 2 \cdot t_1 = 0 + 2 \cdot 2 = 4$

12. Call the nth term of the sequence s_n. Then $s_n = \dfrac{(-1)^n}{n!}$ for all integers $n \geq 0$. So for all integers $k \geq 1$, $s_k = \dfrac{(-1)^k}{k!}$ and $s_{k-1} = \dfrac{(-1)^{k-1}}{(k-1)!}$. It follows that for all integers $k \geq 1$,

$$\frac{-s_{k-1}}{k} = \frac{-\frac{(-1)^{k-1}}{(k-1)!}}{k} = \frac{-(-1)^{k-1}}{k \cdot (k-1)!} = \frac{(-1)^k}{k!} = s_k.$$

18. *b.* $a_4 = 26 + 1 + 26 + 1 + 26 = 80$

21. *a.* $t_1 = 2$, $t_2 = 2 + 2 + 2 = 6$

c. For all integers $k \geq 2$,

$$
\begin{aligned}
t_k \;=\;\; & t_{k-1} && \text{(moves to transfer the top } 2k-2 \text{ disks from pole } A \text{ to pole } B) \\
& +2 && \text{(moves to transfer the bottom two disks from pole } A \text{ to pole } C) \\
& +t_{k-1} && \text{(moves to transfer the top } 2k-2 \text{ disks from pole } B \text{ to pole } C) \\
=\;\; & 2t_{k-1} + 2.
\end{aligned}
$$

Note that transferring the stack of $2k$ disks from pole A to pole C requires at least two transfers of the top $2(k-1)$ disks: one to transfer them off the bottom two disks to free the bottom disks so that they can be moved to pole C and another to transfer the top $2(k-1)$ disks back on top of the bottom two disks. Thus at least $2t_{k-1}$ moves are needed to effect these two transfers. Two more moves are needed to transfer the bottom two disks from pole A to pole C, and this transfer cannot be effected in fewer than two moves. It follows that the sequence of moves indicated in the description of the equation above is, in fact, minimal.

24. $F_{13} = F_{12} + F_{11} = 233 + 144 = 377$, $F_{14} = F_{13} + F_{12} = 377 + 233 = 610$

30. *d. Proof (by mathematical induction):* Let the property $P(n)$ be the equation $F_{n+2}F_n - F_{n+1}^2 = (-1)^n$.

Show that the property is true for $n = 0$: The property is true for $n = 0$ because for $n = 0$ the left-hand side is $F_{0+2}F_0 - F_1^2 = 2 \cdot 1 - 1^2 = 1$, and the right-hand side is $(-1)^0 = 1$ also.

Show that for all integers $k \geq 0$, if the property is true for $n = k$ then it is true for $n = k + 1$: Let k be an integer with $k \geq 0$, and suppose that $F_{k+2}F_k - F_{k+1}^2 = (-1)^k$ for some integer $k \geq 0$. *[This is the inductive hypothesis.]* We must show that $F_{k+3}F_{k+1} - F_{k+2}^2 = (-1)^k$. But by inductive hypothesis,

$$F_{k+1}^2 = F_{k+2}F_k - (-1)^k = F_{k+2}F_k + (-1)^{k+1}. \quad \text{[This is equation (*).]}$$

Hence,

$$
\begin{aligned}
F_{k+3}F_{k+1} &- F_{k+2}^2 \\
&= (F_{k+1} + F_{k+2})F_{k+1} - F_{k+2}^2 && \text{by definition of the Fibonacci sequence} \\
&= F_{k+1}^2 + F_{k+2}F_{k+1} - F_{k+2}^2 \\
&= F_{k+2}F_k + (-1)^{k+1} + F_{k+2}F_{k+1} - F_{k+2}^2 && \text{by substitution from equation (*)} \\
&= F_{k+2}(F_k + F_{k+1} - F_{k+2}) + (-1)^{k+1} && \text{by factoring out } F_{k+2} \\
&= F_{k+2}(F_{k+2} - F_{k+2}) + (-1)^{k+1} && \text{by definition of the Fibonacci sequence} \\
&= F_{k+2} \cdot 0 + (-1)^{k+1} \\
&= (-1)^{k+1}.
\end{aligned}
$$

33. Let $L = \lim_{n \to \infty} x_n$. By definition of $x_0, x_1, x_2, \ldots$ and by the continuity of the square root function,

$$L = \lim_{n \to \infty} x_n = \lim_{n \to \infty} \sqrt{2 + x_{n-1}} = \sqrt{2 + \lim_{n \to \infty} x_{n-1}} = \sqrt{2 + L}.$$

Hence $L^2 = 2 + L$, and so $L^2 - L - 2 = 0$. Factoring gives $(L - 2)(L + 1) = 0$, and so $L = 2$ or $L = -1$. But $L \geq 0$ because each $x_i \geq 0$. Thus $L = 2$.

42. To get a sense of the problem, we compute s_4 directly. If there are four seats in the row, there can be a single student in any one of the four seats or there can be a pair of students in seats 1&3, 1&4, or 2&4. No other arrangements are possible because with more than two students, two would have to sit next to each other. Thus $s_4 = 4 + 3 = 7$. In general, if there are k chairs in a row, then

$$
\begin{aligned}
s_k \;=\; & s_{k-1} && \text{(the number of ways a nonempty set of students can sit} \\
& && \text{in the row with no two students adjacent and chair } k \text{ empty)} \\[4pt]
& +s_{k-2} && \text{(the number of ways students can sit in the row with chair } k \\
& && \text{occupied, chair } k-1 \text{ empty, and chairs 1 through} \\
& && k-2 \text{ occupied by a nonempty set of students in such a} \\
& && \text{way that no two students are adjacent)} \\[4pt]
& +1 && \text{(for the seating in which chair } k \text{ is occupied} \\
& && \text{and all the other chairs are empty)} \\[4pt]
\;=\; & s_{k-1} + s_{k-2} + 1 \text{ for all integers } k \geq 3.
\end{aligned}
$$

51. *Proof (by strong mathematical induction):* Let the property $P(n)$ be the inequality $F_n < 2^n$ where F_n is the nth Fibonacci number.

Show that the property is true for $n = 1$ and $n = 2$: $F_1 = 1 < 2 = 2^1$ and $F_2 = 3 < 4 = 2^2$.

Show that for all integers $k > 2$, if the property is true for all integers i with $1 \leq i < k$ then it is true for k: Let k be an integer with $k > 2$, and suppose that $F_i < 2^i$ for all integers i with $0 \leq i < k$. *[This is the inductive hypothesis.]* We must show that $F_k < 2^k$. Now by definition of the Fibonacci numbers, $F_k = F_{k-1} + F_{k-2}$. But by inductive hypothesis *[since $k > 2$]*, $F_{k-1} < 2^{k-1}$ and $F_{k-2} < 2^{k-2}$. Hence $F_k = F_{k-1} + F_{k-2} < 2^{k-1} + 2^{k-2} = 2^{k-2} \cdot (2 + 1) = 3 \cdot 2^{k-2} < 4 \cdot 2^{k-2} = 2^k$. Thus $F_k < 2^k$ *[as was to be shown]*.

[Since both the basis and inductive steps have been proved, we conclude that $F_n < 2^n$ for all integers $n \geq 1$.]

54. *a.* $d_1 = 0$, $d_2 = 1$, $d_3 = 2$ (231 and 312)

 b. $d_4 = 9$ (2143, 3412, 4321, 3142, 4123, 2413, 4312, 2341, 3421)

 c. Divide the set of all derangements into two subsets: one subset, S, consists of all derangements in which the number 1 changes places with another number, and the other subset, T, consists of all derangements in which the number 1 goes to position $i \neq 1$ but i does not go to position 1. Forming a derangement in S can be regarded as a two-step process: step 1 is to choose a position i and to interchange 1 and i and step 2 is to derange the remaining $k - 2$ numbers. Now there are $k - 1$ numbers with which 1 can trade places in step 1, and so by the product rule there are $(k-1)d_{k-2}$ derangements in S. Forming a derangement in T can also be regarded as a two-step process: step 1 is to derange the $k - 1$ numbers $2, 3, \ldots, k$ in positions $2, 3, \ldots, k$, and step 2 is to interchange the number 1 in position 1 with any of the numbers in the derangement. Now there are $k - 1$ choices of numbers to interchange 1 with in step 2, and so by the multiplication rule there are $d_{k-1}(k - 1)$ derangements in T. It follows that the total number of derangements of the given k numbers is $d_k = (k - 1)d_{k-1} + (k - 1)d_{k-2}$ for all integers $k \geq 3$.

Section 8.2

6.
$$d_1 = 2$$
$$d_2 = 2d_1 + 3 = 2 \cdot 2 + 3 = 2^2 + 3$$
$$d_3 = 2d_2 + 3 = 2(2^2 + 3) + 3 = 2^3 + 2 \cdot 3 + 3$$
$$d_4 = 2d_3 + 3 = 2(2^3 + 2 \cdot 3 + 3) + 3 = 2^4 + 2^2 \cdot 3 + 2 \cdot 3 + 3$$
$$d_5 = 2d_4 + 3 = 2(2^4 + 2^2 \cdot 3 + 2 \cdot 3 + 3) + 3 = 2^5 + 2^3 \cdot 3 + 2^2 \cdot 3 + 2 \cdot 3 + 3$$

$$\vdots$$

$$
\begin{aligned}
\text{Guess: } d_n &= 2^n + 2^{n-2} \cdot 3 + 2^{n-3} \cdot 3 + \cdots + 2^2 \cdot 3 + 2 \cdot 3 + 3 \\
&= 2^n + 3(2^{n-2} + 2^{n-3} + \cdots + 2^2 + 2 + 1) \\
&= 2^n + 3\left(\frac{2^{(n-2)+1} - 1}{2 - 1}\right) \quad \text{[by Theorem 4.2.3]} \\
&= 2^n + 3(2^{n-1} - 1) \\
&= 2^{n-1}(2 + 3) - 3 = 5 \cdot 2^{n-1} - 3 \quad \text{for all integers } n \geq 1
\end{aligned}
$$

9.
$$g_1 = 1$$
$$g_2 = \frac{g_1}{g_1 + 2} = \frac{1}{1 + 2} = \frac{1}{1 + 2}$$
$$g_3 = \frac{g_2}{g_2 + 2} = \frac{\frac{1}{1+2}}{\frac{1}{1+2} + 2} = \frac{1}{1 + 2(1 + 2)} = \frac{1}{1 + 2 + 2^2}$$
$$g_4 = \frac{g_3}{g_3 + 2} = \frac{\frac{1}{1+2+2^2}}{\frac{1}{1+2+2^2} + 2} = \frac{1}{1 + 2(1 + 2 + 2^2)} = \frac{1}{1 + 2 + 2^2 + 2^3}$$
$$g_5 = \frac{g_4}{g_4 + 2} = \frac{\frac{1}{1+2+2^2+2^3}}{\frac{1}{2(1+2+2^2+2^3)} + 2} = \frac{1}{1 + 2 + 2^2 + 2^3 + 2^4}$$

$$\vdots$$

$$\text{Guess: } g_n = \frac{1}{1 + 2 + 2^2 + 2^3 + \cdots + 2^{n-1}}$$

$$= \frac{1}{2^n - 1} \quad \textit{[by Theorem 4.2.3)]} \quad \text{for all integers } n \geq 1$$

15.

$$
\begin{aligned}
y_1 &= 1 \\
y_2 &= y_1 + 2^2 = 1 + 2^2 \\
y_3 &= y_2 + 3^2 = (1 + 2^2) + 3^2 = 1 + 2^2 + 3^2 \\
y_4 &= y_3 + 4^2 = (1 + 2^2 + 3^2) + 4^2 = 1^2 + 2^2 + 3^2 + 4^2
\end{aligned}
$$

⋮

Guess:

$$y_n = 1^2 + 2^2 + 3^2 + \cdots + n^2 = \frac{n(n+1)(2n+1)}{6} \quad \text{by exercise 10 of Section 4.2}$$

21. *Proof*: Let r be a fixed constant and $a_0, a_1, a_2, \ldots$ a sequence that satisfies the recurrence relation $a_k = ra_{k-1}$ for all integers $k \geq 1$ and the initial condition $a_0 = a$. Let the property $P(n)$ be the equation $a_n = ar^n$.

Show that the property is true for $n = 0$: For $n = 0$ the right-hand side of the equation is $ar^0 = a \cdot 1 = a$, which is also the left-hand side of the equation.

Show that for all integers $k \geq 0$, if the property is true for $n = k$ then it is true for $n = k + 1$: Let k be an integer with $k \geq 0$, and suppose that $a_k = ar^k$. *[This is the inductive hypothesis.]* We must show that $a_{k+1} = ar^{k+1}$. But

$$
\begin{aligned}
a_{k+1} &= ra_k && \text{by definition of } a_0, a_1, a_2, \ldots \\
&= r(ar^k) && \text{by substitution from the inductive hypothesis} \\
&= ar^{k+1} && \text{by the laws of exponents.}
\end{aligned}
$$

[This is what was to be shown.]

27. *a.* Let the original balance in the account be A dollars, and let A_n be the amount owed in month n assuming the balance is not reduced by making payments during the year. The annual interest rate is 18%, and so the monthly interest rate is $(18/12)\% = 1.5\% = 0.015$. The sequence $A_0, A_1, A_2, \ldots$ satisfies the recurrence relation $A_k = A_{k-1} + 0.015A_{k-1} = 1.015A_{k-1}$. Thus $A_1 = 1.015A_0 = 1.015A$, $A_2 = 1.015A_1 = 1.015(1.015A) = (1.015)^2A$, ..., $A_{12} = 1.015A_{11} = 1.015(1.015)^{11}A = (1.015)^{12}A$. So the amount owed at the end of the year is $(1.015)^{12}A$. It follows that the APR is $\dfrac{(1.015)^{12}A - A}{A} = \dfrac{A((1.015)^{12} - 1)}{A} = (1.015)^{12} - 1 \cong$ 19.6%.

Note: Because $A_k = 1.015A_{k-1}$ for each integer $k \geq 1$, we could have immediately concluded that the sequence is geometric and, therefore, satisfies the equation $A_n = A_0(1.015)^n = A(1.015)^n$.

b. Because the person pays \$150 per month to pay off the loan, the balance at the end of month k is $B_k = 1.015B_{k-1} - 150$. We use iteration to find an explicit formula for $B_0, B_1, B_2, \ldots$.

$$
\begin{aligned}
B_0 &= 3000 \\
B_1 &= (1.015)B_0 - 150 = 1.015(3000) - 150 \\
B_2 &= (1.015)B_1 - 150 = (1.015)[1.015(3000) - 150] - 150 \\
&= 3000(1.015)^2 - 150(1.015) - 150 \\
B_3 &= (1.015)B_2 - 150 = (1.015)[3000(1.015)^2 - 150(1.015) - 150] - 150 \\
&= 3000(1.015)^3 - 150(1.015)^2 - 150(1.015) - 150 \\
B_4 &= (1.015)B_3 - 150 \\
&= (1.015)[3000(1.015)^3 - 150(1.015)^2 - 150(1.015) - 150] - 150 \\
&= 3000(1.015)^4 - 150(1.015)^3 - 150(1.015)^2 - 150(1.015) - 150
\end{aligned}
$$

Guess: $B_n = 3000(1.015)^n + [150(1.015)^{n-1} - 150(1.015)^{n-2} + \ldots$
$$-150(1.015)^2 - 150(1.015) - 150]$$

$$= 3000(1.015)^n - 150[(1.015)^{n-1} + (1.015)^{n-2} + \cdots + (1.015)^2 + 1.015 + 1]$$

$$= 3000(1.015)^n - 150\left(\frac{(1.015)^n - 1}{1.015 - 1}\right)$$

$$= (1.015)^n(3000) - \frac{150}{0.015}((1.015)^n - 1)$$

$$= (1.015)^n(3000) - 10000((1.015)^n - 1)$$

$$= (1.015)^n(3000 - 10000) + 10000$$

$$= (-7000)(1.015)^n + 10000$$

So it appears that $B_n = (-7000)(1.015)^n + 10000$. We use mathematical induction to confirm this guess.

Proof (by mathematical induction): Let $B_0, B_1, B_2, \ldots$ be a sequence that satisfies the recurrence relation $B_k = (1.015)B_{k-1} - 150$ for all integers $k \geq 1$, with initial condition $B_0 = 3000$, and let the property $P(n)$ be the equation $B_n = (-7000)(1.015)^n + 10000$.

Show that the property is true for $n = 0$: For $n = 0$ the right-hand side of the equation is $(-7000)(1.015)^n + 10000 = 3000$, which equals B_0, the left-hand side of the equation

Show that for all integers $k \geq 0$, if the property is true for $n = k$ then it is true for $n = k + 1$: Let k be an integer with $k \geq 0$, and suppose that $B_k = (-7000)(1.015)^k + 10000$. *[This is the inductive hypothesis.]* We must show that $B_{k+1} = (-7000)(1.015)^{k+1} + 10000$. But

$$
\begin{aligned}
B_{k+1} &= (1.015)B_k - 150 &&\text{by definition of } B_0, B_1, B_2, \ldots \\
&= (1.015)[(-7000)(1.015)^k + 10000] - 150 &&\text{by substitution from} \\
& &&\text{the inductive hypothesis} \\
&= (-7000)(1.015)^{k+1} + 10150 - 150 \\
&= (-7000)(1.015)^{k+1} + 10000 &&\text{by the laws of algebra.}
\end{aligned}
$$

[This is what was to be shown.]

c. By part (b), $B_n = (-7000)(1.015)^n + 10000$, and so we need to find the value of n for which

$$(-7000)(1.015)^n + 10000 = 0.$$

But this equation holds

$$\Leftrightarrow \quad 7000(1.015)^n = 10000$$

$$\Leftrightarrow \quad (1.015)^n = \frac{10000}{7000} = \frac{10}{7}$$

$$\Leftrightarrow \quad \log_{10}(1.015)^n = \log_{10}\left(\frac{10}{7}\right) \quad \text{by property (7.2.5)}$$

$$\Leftrightarrow \quad n \log_{10}(1.015) = \log_{10}\left(\frac{10}{7}\right) \quad \text{by exercise 31 of Section 7.2}$$

$$\Leftrightarrow \quad n = \frac{\log_{10}(10/7)}{\log_{10}(1.015)} \cong 24.$$

So $n \cong 24$ months $= 2$ years. It will require approximately 2 years to pay off the balance, assuming that payments of \$150 are made each month and the balance is not increased by any additional purchases.

d. Assuming that the person makes no additional purchases and pays \$150 each month, the person will have made 24 payments of \$150 each, for a total of \$3600 to pay off the initial balance of \$3000.

33. *Proof (by mathematical induction):* Let $f_1, f_2, f_3, \ldots$ be a sequence that satisfies the recurrence relation $f_k = f_{k-1} + 2^k$ for all integers $k \geq 2$, with initial condition $f_1 = 1$, and let the property $P(n)$ be the equation $f_n = 2^{n+1} - 3$.

Show that the property is true for $n = 1$: For $n = 1$ the right-hand side of the equation is $2^{1+1} - 3 = 4 - 3 = 1$, which equals f_1, the left-hand side of the equation.

Show that for all integers $k \geq 1$, if the property is true for $n = k$ then it is true for $n = k + 1$: Let k be an integer with $k \geq 1$, and suppose that $f_k = 2^{k+1} - 3$. *[This is the inductive hypothesis.]* We must show that $f_{k+1} = 2^{(k+1)+1} - 3$, or, equivalently, that $f_{k+1} = 2^{k+2} - 3$. But

$$\begin{aligned} f_{k+1} &= f_k + 2^{k+1} & \text{by definition of } f_1, f_2, f_3, \ldots \\ &= 2^{k+1} - 3 + 2^{k+1} & \text{by substitution from the inductive hypothesis} \\ &= 2 \cdot 2^{k+1} - 3 \\ &= 2^{k+2} - 3 & \text{by the laws of algebra.} \end{aligned}$$

[This is what was to be shown.]

36. *Proof by mathematical induction:* Let $p_1, p_2, p_3, \ldots$ be a sequence that satisfies the recurrence relation $p_k = p_{k-1} + 2 \cdot 3^k$ for all integers $k \geq 2$, with initial condition $p_1 = 2$, and let the property $P(n)$ be the equation $p_n = 3^{n+1} - 7$.

Show that the property is true for $n = 1$: For $n = 1$ the right-hand side of the equation is $3^{1+1} - 7 = 3^2 - 7 = 9 - 7 = 2$, which equals p_1, the left-hand side of the equation.

Show that for all integers $k \geq 1$, if the property is true for $n = k$ then it is true for $n = k + 1$: Let k be an integer with $k \geq 1$, and suppose that $p_k = 3^{k+1} - 7$. *[This is the inductive hypothesis.]* We must show that $p_{k+1} = 3^{(k+1)+1} - 7 = 3^{k+2} - 7$. But

$$\begin{aligned} p_{k+1} &= p_k + 2 \cdot 3^{k+1} & \text{by definition of } p_1, p_2, p_3, \ldots \\ &= (3^{k+1} - 7) + 2 \cdot 3^{k+1} & \text{by substitution from the inductive hypothesis} \\ &= 3^{k+1}(1 + 2) - 7 \\ &= 3 \cdot 3^{k+1} - 7 \\ &= 3^{k+2} - 7 & \text{by the laws of algebra.} \end{aligned}$$

[This is what was to be shown.]

42. *Proof by mathematical induction:* Let $t_1, t_2, t_3, \ldots$ be a sequence that satisfies the recurrence relation $t_k = 2t_{k-1} + 2$ for all integers $k \geq 2$, with initial condition $t_1 = 2$, and let the property $P(n)$ be the equation $t_n = 2^{n+1} - 2$.

Show that the property is true for $n = 1$ For $n = 1$ the right-hand side of the equation is $t_1 = 2^2 - 2 = 2$, which is true.

Show that for all integers $k \geq 1$, if the property is true for $n = k$ then it is true for $n = k + 1$: Let k be an integer with $k \geq 1$, and suppose that $t_k = 2^{k+1} - 2$. *[This is the inductive hypothesis.]* We must show that $t_{k+1} = 2^{(k+1)+1} - 2$, or, equivalently, $t_{k+1} = 2^{k+2} - 2$. But

$$\begin{aligned} t_{k+1} &= 2t_k + 2 & \text{by definition of } t_1, t_2, t_3, \ldots \\ &= 2(2^{k+1} - 2) + 2 & \text{by inductive hypothesis} \\ &= 2^{k+2} - 4 + 2 \\ &= 2^{k+2} - 2 \end{aligned}$$

[This is what was to be shown.]

48. *a.*

$$w_1 = 1$$
$$w_2 = 2$$
$$w_3 = w_1 + 3 = 1 + 3$$
$$w_4 = w_2 + 4 = 2 + 4$$
$$w_5 = w_3 + 5 = 1 + 3 + 5$$
$$w_6 = w_4 + 6 = 2 + 4 + 6$$
$$w_7 = w_5 + 7 = 1 + 3 + 5 + 7$$

Guess:

$$w_n = \begin{cases} 1 + 3 + 5 + \cdots + n & \text{if } n \text{ is odd} \\ 2 + 4 + 6 + \cdots + n & \text{if } n \text{ is even} \end{cases}$$

$$= \begin{cases} \left(\dfrac{n+1}{2}\right)^2 & \text{if } n \text{ is odd} \\ 2\left(1 + 2 + 3 + \ldots + \dfrac{n}{2}\right) & \text{if } n \text{ is even} \end{cases} \quad \text{by exercise 5 of Section 4.2}$$

$$= \begin{cases} \left(\dfrac{n+1}{2}\right)^2 & \text{if } n \text{ is odd} \\ 2\left(\dfrac{\dfrac{n}{2}\left(\dfrac{n}{2} + 1\right)}{2}\right) & \text{if } n \text{ is even} \end{cases} \quad \text{by Theorem 4.2.1}$$

$$= \begin{cases} \dfrac{(n+1)^2}{4} & \text{if } n \text{ is odd} \\ \dfrac{n(n+1)}{4} & \text{if } n \text{ is even} \end{cases} \quad \text{by the laws of algebra.}$$

b. Proof by strong mathematical induction: Let $w_1, w_2, w_3, \ldots$ be a sequence that satisfies the recurrence relation $w_k = w_{k-2} + k$ for all integers $k \geq 2$, with initial conditions $w_1 = 1$ and $w_2 = 2$, and let the property $P(n)$ be the equation

$$w_n = \begin{cases} \dfrac{(n+1)^2}{4} & \text{if } n \text{ is odd} \\ \dfrac{n(n+2)}{4} & \text{if } n \text{ is even} \end{cases} \quad \text{for all integers } n \geq 1.$$

Show that the property is true for $n = 1$ and $n = 2$: For $n = 1$ and $n = 2$ the right-hand sides of the equation are $(1+1)^2/4 = 1$ and $2(2+2)/4 = 2$, which equal w_1 and w_2 respectively.

Show that for all integers $k > 2$, if the property is true for all i with $1 \leq i < k$ then it is true for k: Let k be an integer with $k > 2$ and suppose

$$w_i = \begin{cases} \dfrac{(i+1)^2}{4} & \text{if } i \text{ is odd} \\ \dfrac{i(i+2)}{4} & \text{if } i \text{ is even} \end{cases} \quad \text{for all integers } i \text{ with } 1 \leq i < k.$$

[*This is the inductive hypothesis.*]

We must show that

$$w_k = \begin{cases} \dfrac{(k+1)^2}{4} & \text{if } k \text{ is odd} \\ \dfrac{k(k+2)}{4} & \text{if } k \text{ is even.} \end{cases}$$

But

$$w_k = w_{k-2} + k \qquad\qquad \text{by definition of } w_1, w_2, w_3, \ldots$$

$$= \begin{cases} \dfrac{((k-2)+1)^2}{4} + k & \text{if } k-2 \text{ is odd} \\[2mm] \dfrac{(k-2)((k-2)+2)}{4} + k & \text{if } k-2 \text{ is even} \end{cases} \qquad \text{by substitution from the inductive hypothesis}$$

$$= \begin{cases} \dfrac{(k-1)^2}{4} + \dfrac{4k}{4} & \text{if } k \text{ is odd} \\[2mm] \dfrac{(k-2)\cdot k}{4} + \dfrac{4k}{4} & \text{if } k \text{ is even} \end{cases} \qquad \text{because } k-2 \text{ and } k \text{ have the same parity}$$

$$= \begin{cases} \dfrac{k^2 - 2k + 1 + 4k}{4} & \text{if } k \text{ is odd} \\[2mm] \dfrac{k^2 - 2k + 4k}{4} & \text{if } k \text{ is even} \end{cases}$$

$$= \begin{cases} \dfrac{k^2 + 2k + 1}{4} & \text{if } k \text{ is odd} \\[2mm] \dfrac{k^2 + 2k}{4} & \text{if } k \text{ is even} \end{cases}$$

$$= \begin{cases} \dfrac{(k+1)^2}{4} & \text{if } k \text{ is odd} \\[2mm] \dfrac{k(k+2)}{4} & \text{if } k \text{ is even} \end{cases} \qquad \text{by the laws of algebra.}$$

[This is what was to be shown.]

51. The sequence does not satisfy the formula. By definition of $a_1, a_2, a_3, \ldots$, $a_1 = 0$, $a_2 = (a_1 + 1)^2 = 1^2 = 1$, $a_3 = (a_2 + 1)^2 = (1+1)^2 = 4$, $a_4 = (a_3 + 1)^2 = (4+1)^2 = 25$. But according to the formula $a_4 = (4-1)^2 = 9 \neq 25$.

54. *a.*

$$Y_1 = E + c + mY_0$$

$$Y_2 = E + c + mY_1 = E + c + m(E + c + mY_0) = (E + c) + m(E + c) + m^2 Y_0$$

$$Y_3 = E + c + mY_2 = E + c + m((E+c) + m(E+c) + m^2 Y_0) = (E+c) + m(E+c) + m^2(E+c) + m^3 Y_0$$

$$Y_4 = E + c + mY_3 = E + c + m((E+c) + m(E+c) + m^2(E+c) + m^3 Y_0)$$

$$= (E+c) + m(E+c) + m^2(E+c) + m^3(E+c) + m^4 Y_0$$

$$\text{Guess: } Y_n = (E+c) + m(E+c) + m^2(E+c) + \cdots + m^{n-1}(E+c) + m^n Y_0$$

$$= (E+c)[1 + m + m^2 + \cdots + m^{n-1}] + m^n Y_0$$

$$= (E+c)\left(\frac{m^n - 1}{m-1}\right) + m^n Y_0, \text{ for all integers } n \geq 1.$$

b. Suppose $0 < m < 1$. Then

$$\lim_{n \to \infty} Y_n = \lim_{n \to \infty} \left((E + c) \left(\frac{m^n - 1}{m - 1} \right) + m^n Y_0 \right)$$

$$= (E + c) \left(\frac{\lim_{n \to \infty} m^n - 1}{m - 1} \right) + \lim_{n \to \infty} m^n Y_0$$

$$= (E + c) \left(\frac{0 - 1}{m - 1} \right) + 0 \cdot Y_0 \qquad \text{because when } 0 < m < 1,$$

$$\text{then } \lim_{n \to \infty} m^n = 0$$

$$= \frac{E + c}{1 - m}.$$

Section 8.3

3. *b.*

$$\left\{ \begin{array}{l} a_0 = C \cdot 2^0 + D = C + D = 0 \\ a_1 = C \cdot 2^1 + D = 2C + D = 2 \end{array} \right\} \Leftrightarrow \left\{ \begin{array}{l} D = -C \\ 2C + (-C) = 2 \end{array} \right\} \Leftrightarrow \left\{ \begin{array}{l} D = -C \\ C = 2 \end{array} \right\} \Leftrightarrow \left\{ \begin{array}{l} C = 2 \\ D = -2 \end{array} \right\}$$

$$a_2 = C \cdot 2^2 + D = 2 \cdot 2^2 + (-2) = 6$$

6. *Proof*: Given that $b_n = C \cdot 3^n + D(-2)^n$, then for any choice of C and D and integer $k \geq 2$, $b_k = C \cdot 3^k + D(-2)^k$, $b_{k-1} = C \cdot 3^{k-1} + D(-2)^{k-1}$, and $b_{k-2} = C \cdot 3^{k-2} + D(-2)^{k-2}$. Hence, $b_{k-1} + 6b_{k-2} = \left(C \cdot 3^{k-1} + D(-2)^{k-1} \right) + 6 \left(C \cdot 3^{k-2} + D(-2)^{k-2} \right) = C \cdot \left(3^{k-1} + 6 \cdot 3^{k-2} \right) + D \left((-2)^{k-1} + 6(-2)^{k-2} \right) = C \cdot 3^{k-2} (3 + 6) + D(-2)^{k-2} (-2 + 6) = C \cdot 3^{k-2} \cdot 3^2 + D(-2)^{k-2} 2^2 = C \cdot 3^k + D(-2)^k = b_k$.

9. *a.* If for all integers $k \geq 2$, $t^k = 7t^{k-1} - 10t^{k-2}$ and $t \neq 0$, then $t^2 = 7t - 10$ and so $t^2 - 7t + 10 = 0$. But $t^2 - 7t + 10 = (t - 2)(t - 5)$. Thus $t = 2$ or $t = 5$.

b. It follows from part (a) and the distinct roots theorem that for some constants C and D, $b_0, b_1, b_2, \ldots$ satisfies the equation $b_n = C \cdot 2^n + D \cdot 5^n$ for all integers $n \geq 0$. Since $b_0 = 2$ and $b_1 = 2$, then

$$\left\{ \begin{array}{l} b_0 = C \cdot 2^0 + D \cdot 5^0 = C + D = 2 \\ b_1 = C \cdot 2^1 + D \cdot 5^1 = 2C + 5D = 2 \end{array} \right\} \Leftrightarrow \left\{ \begin{array}{l} D = 2 - C \\ 2C + 5(2 - C) = 2 \end{array} \right\}$$

$$\Leftrightarrow \left\{ \begin{array}{l} D = 2 - C \\ C = 8/3 \end{array} \right\} \qquad \Leftrightarrow \left\{ \begin{array}{l} D = 2 - (8/3) = -(2/3) \\ C = 8/3 \end{array} \right\}$$

Thus $b_n = \dfrac{8}{3} \cdot 2^n - \dfrac{2}{3} \cdot 5^n$ for all integers $n \geq 0$.

12. The characteristic equation is $t^2 - 9 = 0$. Since $t^2 - 9 = (t - 3)(t + 3)$, the roots are $t = 3$ and $t = -3$. By the distinct roots theorem, for some constants C and D, $e_n = C \cdot 3^n + D(-3)^n$ for all integers $n \geq 0$. Since $e_0 = 0$ and $e_1 = 2$, then

$$\left\{ \begin{array}{l} e_0 = C \cdot 3^0 + D(-3)^0 = C + D = 0 \\ e_1 = C \cdot 3^1 + D(-3)^1 = 3C - 3D = 2 \end{array} \right\} \Leftrightarrow \left\{ \begin{array}{l} D = -C \\ 3C - 3(-C) = 2 \end{array} \right\} \Leftrightarrow \left\{ \begin{array}{l} D = -1/3 \\ C = 1/3 \end{array} \right\}$$

Thus $e_n = \dfrac{1}{3} \cdot 3^n - \dfrac{1}{3}(-3)^n = 3^{n-1} + (-3)^{n-1} = 3^{n-1}(1 + (-1)^{n-1}) = \left\{ \begin{array}{ll} 2 \cdot 3^{n-1} & \text{if } n \text{ is odd} \\ 0 & \text{if } n \text{ is even} \end{array} \right.$

for all integers $n \geq 0$.

15. The characteristic equation is $t^2 - 6t + 9 = 0$. Since $t^2 - 6t + 9 = (t-3)^2$, there is only one root, $t = 3$. By the single root theorem, for some constants C and D, $t_n = C \cdot 3^n + D \cdot n \cdot 3^n$ for all integers $n \geq 0$. Since $t_0 = 1$ and $t_1 = 3$, then

$$\left\{ \begin{array}{l} t_0 = C \cdot 3^0 + D \cdot 0 \cdot 3^0 = C = 1 \\ t_1 = C \cdot 3^1 + D \cdot 1 \cdot 3^1 = 3C + 3D = 3 \end{array} \right\} \Leftrightarrow \left\{ \begin{array}{l} C = 1 \\ C + D = 1 \end{array} \right\} \Leftrightarrow \left\{ \begin{array}{l} C = 1 \\ D = 0 \end{array} \right\}$$

Thus $t_n = 1 \cdot 3^n + 0 \cdot n \cdot 3^n = 3^n$ for all integers $n \geq 0$.

18. *Proof*: Suppose that $s_0, s_1, s_2, \ldots$ and $t_0, t_1, t_2, \ldots$ are sequences such that $s_k = 5s_{k-1} - 4s_{k-2}$ and $t_k = 5t_{k-1} - 4t_{k-2}$ for all integers $k \geq 2$. Then for all integers $k \geq 2$, $5(2s_{k-1} + 3t_{k-1}) - 4(2s_{k-2} + 3t_{k-2}) = (5 \cdot 2s_{k-1} - 4 \cdot 2s_{k-2}) + (5 \cdot 3t_{k-1} - 4 \cdot 3t_{k-2}) = 2(5s_{k-1} - 4s_{k-2}) + 3(5t_{k-1} - 4t_{k-2}) = 2s_k + 3t_k$. This is what was to be shown.

21. Let $a_0, a_1, a_2, \ldots$ be any sequence that satisfies the recurrence relation $a_k = Aa_{k-1} + Ba_{k-2}$ for some real numbers A and B with $B \neq 0$ and for all integers $k \geq 2$. Furthermore, suppose that the equation $t^2 - At - B = 0$ has a single real root r. First note that $r \neq 0$ because otherwise we would have $0^2 - A \cdot 0 - B = 0$, which would imply that $B = 0$ and contradict the hypothesis. Second, note that the following system of equations with unknowns C and D has a unique solution.

$$a_0 = Cr^0 + 0 \cdot Dr^0 = 1 \cdot C + 0 \cdot D$$
$$a_1 = Cr^1 + 1 \cdot Dr^1 = C \cdot r + D \cdot r$$

One way to reach this conclusion is to observe that the determinant of the system is $1 \cdot r - r \cdot 0 = r \neq 0$. Another way to reach the conclusion is to write the system as

$$a_0 = C$$
$$a_1 = Cr + Dr$$

and let $C = a_0$ and $D = (a_1 - Cr)/r$. It is clear by substitution that these values of C and D satisfy the system. Conversely, if any numbers C and D satisfy the system, then $C = a_0$ and substituting C into the second equation and solving for D yields $D = (a_1 - Cr)/r$.

Proof of the exercise statement by strong mathematical induction: Let $a_0, a_1, a_2, \ldots$ be any sequence that satisfies the recurrence relation $a_k = Aa_{k-1} + Ba_{k-2}$ for some real numbers A and B with $B \neq 0$ and for all integers $k \geq 2$. Furthermore, suppose that the equation $t^2 - At - B = 0$ has a single real root r. Let the property $P(n)$ be the equation $a_n = Cr^n + nDr^n$ where C and D are the unique real numbers such that $a_0 = Cr^0 + 0 \cdot Dr^0$ and $a_1 = Cr^1 + 1 \cdot Dr^1$.

Show that the property is true for $n = 0$ and $n = 1$: The fact that the property is true for $n = 0$ and $n = 1$ is automatic because C and D are exactly those numbers for which $a_0 = Cr^0 + 0 \cdot Dr^0$ and $a_1 = Cr^1 + 1 \cdot Dr^1$.

Show that for all integers $k > 1$, if the property is true for $n = k$ then it is true for $n = k + 1$: Let k be an integer with $k > 1$ and suppose $a_i = Cr^i + iDr^i$, for all integers i with $1 \leq i < k$. *[This is the inductive hypothesis.]* We must show that $a_k = Cr^k + kDr^k$. Now by the inductive hypothesis, $a_{k-1} = Cr^{k-1} + (k-1)Dr^{k-1}$ and $a_{k-2} = Cr^{k-2} + (k-2)Dr^{k-2}$. So

$$\begin{aligned} a_k &= Aa_{k-1} + Ba_{k-2} && \text{by definition of } a_0, a_1, a_2, \ldots \\ &= A(Cr^{k-1} + (k-1)Dr^{k-1}) + B(Cr^{k-2} + (k-2)Dr^{k-2}) \\ && \text{by substitution from the inductive hypothesis} \\ &= C(Ar^{k-1} + Br^{k-2}) + D(A(k-1)r^{k-1} + B(k-2)r^{k-2}) \\ && \text{by algebra} \\ &= Cr^k + Dkr^k && \text{by Lemma 8.3.4.} \end{aligned}$$

[This is what was to be shown.]

24. *a.* If $\dfrac{\phi}{1} = \dfrac{1}{\phi - 1}$, then $\phi(\phi - 1) = 1$, or, equivalently, $\phi^2 - \phi - 1 = 0$ and so ϕ satisfies the equation $t^2 - t - 1 = 0$.

b. By the quadratic formula, the solutions to $t^2 - t - 1 = 0$ are $t = \dfrac{1 \pm \sqrt{1 + 4}}{2} = \begin{cases} (1 + \sqrt{5})/2 \\ (1 - \sqrt{5})/2 \end{cases}$

Let $\phi_1 = (1 + \sqrt{5})/2$ and $\phi_2 = (1 - \sqrt{5})/2$

c. $F_n = \dfrac{1}{\sqrt{5}} \cdot \phi_1^{n+1} - \dfrac{1}{\sqrt{5}} \cdot \phi_2^{n+1} = \dfrac{1}{\sqrt{5}}(\phi_1^{n+1} - \phi_2^{n+1})$

This equation is an alternative way to write equation (8.3.8).

Section 8.4

3. *b.* (1) MI is in the MIU-system by I.

(2) MII is in the MIU-system by (1) and $II(b)$.

(3) $MIIII$ is in the MIU-system by (2) and $II(b)$.

(4) $MIIIIIIII$ is in the MIU-system by (3) and $II(b)$.

(5) $MUIIIII$ is in the MIU-system by (4) and $II(c)$.

(6) $MUIIU$ is in the MIU-system by (5) and $II(c)$.

6. *b.* Even though the number of its left parentheses equals the number of its right parentheses, this structure is not in P either. Roughly speaking, the reason is that given any parenthesis structure derived from the base structure by repeated application of the rules of the recursion, as you move from left to right along the structure, the total of right parentheses you encounter will never be larger than the number of left parentheses you have already passed by. But if you move along $(( )( ))( )$ from left to right, you encounter an extra right parenthesis in the seventh position.

More formally: Let A be the set of all finite sequences of integers and define a function $g: P \to A$ as follows: for each parenthesis structure S in P, let $g[S] = (a_1, a_2, \ldots, a_n)$ where a_i is the number of left parentheses in S minus the number of right parentheses in S counting from left to right through position i. For instance, if $S = (( )( ))$, then $g[S] = (1, 2, 1, 2, 1, 0)$. By the same argument as in part (a), the final component in $g[S]$ will always be 0. We claim that for all parenthesis structures S in P, each component of $g[S]$ is nonnegative. It follows from the claim that if $(( )( ))( )$ were in P, then all components of $g[(( )( ))( )]$ would be nonnegative. But $g[(( )( ))( )] = (1, 2, 1, 2, 1, 0, -1, 0, 1, 0)$, and one of these components is negative. Consequently, $(( )( ))( )$ is not in P.

Proof of the claim (by structural induction): Let the property be the following sentence: Each component of $g[S]$ is nonnegative, with the final component being 0.

Show that each object in the BASE for P satisfies the property: The only object in the base is ( ), and $g[( )] = (1, 0)$. Both components of $g[( )]$ are nonnegative, and the final component is 0.

Show that for each rule in the RECURSION for P, if the rule is applied to objects in P that satisfy the property, then the objects defined by the rule also satisfy the property: The recursion for P consists of two rules, denoted II(a) and II(b). Let S and T be parenthesis structures in P with the property that all components of $g[S]$ and $g[T]$ are nonnegative, with the final components of both S and T being 0. Consider the effect of applying rules II(a) and II(b).

In case rule II(a) is applied to S, the result is (S). Observe that that the first component in $g[(S)]$ is 1 (because (S) starts with a left parenthesis), every subsequent component of $g[(S)]$

except the last is one more than a corresponding (nonnegative) component of $g[S]$. Thus the next-to-last component of (S) is 1 *[because (S) starts with a left parenthesis]*, and the final right-parenthesis of (S) reduces the final component of $g[(S)]$ to 0 also. So each component of $g[(S)]$ is nonnegative, and the final component is 0.

In case rule II(b) is applied to S and T, the result is ST. We must show that each component of $g[ST]$ is nonnegative, with the final component being 0. For concreteness, suppose $g[S]$ has m components and $g[T]$ has n components. The first through the mth components of $g[ST]$ are the same as the first through the mth components of S, which are all nonnegative, with the mth component being 0. Thus the $(m+1)$st through the $(m+n)$th components of $g[ST]$ are the same as the first through the nth components of T, which are all nonnegative, with the final component being 0. Because the final component of $g[ST]$ is the same as the final component of T, all components of $g[ST]$ are nonnegative, with the final component being 0.

By the restriction condition, there are no other elements of P besides those obtainable from the base and recursion conditions. Hence, for all S in P, all components of $g[S]$ are nonnegative, with final component 0.

9. *Proof (by structural induction)*: Let the property be the following sentence: The string begins with an a.

   Show that each object in the BASE for S satisfies the property: The only object in the base is a, and the string a begins with an a.

   Show that for each rule in the RECURSION for S, if the rule is applied to objects in S that satisfy the property, then the objects defined by the rule also satisfy the property: The recursion for S consists of two rules, denoted II(a) and II(b). In case rule II(a) is applied to a string s in S that begins with a a, the result is the string sa, which begins with the same character as s, namely a. Similarly, in case rule II(b) is applied to a string s that begins with a a, the result is the string sb, which also begins with an a. Thus, when each rule in the RECURSION is applied to strings in S that begin with an a, the results are also strings that begin with an a. Because no objects other than those obtained through the BASE and RECURSION conditions are contained in S, every string in S begins with an a.

12. *Proof (by structural induction)*: Let the property be the following sentence: The string represents an odd integer.

   Show that each object in the BASE for S satisfies the property: The objects in the base are 1, 3, 5, 7, and 9. All of these strings represent odd integers.

   Show that for each rule in the RECURSION for S, if the rule is applied to objects in S that satisfy the property, then the objects defined by the rule also satisfy the property: The recursion for S consists of five rules, denoted II(a)–II(e). Suppose s and t are strings in S that represent odd integers. Then the right-most character for each of s and t is 1, 3, 5, 7, or 9. In case rule II(a) is applied to s and t, the result is the string st, which has the same right-most character as t. So st represents an odd integer. In case rules II(b)–II(e) are applied to s, the results are $2s$, $4s$, $6s$, or $8s$. All of these strings have the same right-most character as s, and, therefore, they all represent odd integers. Thus when each rule in the RECURSION is applied to strings in S that represent odd integers, the result is also a string that represents an odd integer. Because no objects other than those obtained through the BASE and RECURSION conditions are contained in S, all the strings in S represent odd integers.

18. Let S be the set of all strings of a's and b's that contain exactly one a. The following is a recursive definition of S.

   I. BASE: $a \in S$

   II. RECURSION: If $s \in S$, then   a. $bs \in S$ b. $sb \in S$

   III. RESTRICTION: There are no elements of S other than those obtained from I and II.

21. *Proof (by mathematical induction)*: Let the property be the sentence "If $a_1, a_2, \ldots, a_n$ and c are any real numbers, then $\prod_{i=1}^{n}(ca_i) = c^n \left(\prod_{i=1}^{n} a_i\right)$."

Show that the property is true for $n = 1$: Let c and a_1 be any real numbers. By the recursive definition of product, both $\prod_{i=1}^{1}(ca_i)$ and $c^1 \prod_{i=1}^{1} a_i$ equal ca_1, and so the property is true for $n = 1$.

Show that for all integers $k \geq 1$, if the property is true for $n = k$ then it is true for $n = k + 1$: Let k be an integer such that $k \geq 1$. Suppose that if c and $a_1, a_2, \ldots, a_k$ are any real numbers, then $\prod_{i=1}^{k}(ca_i) = c^k \left(\prod_{i=1}^{k} a_i\right)$. *[This is the inductive hypothesis.]* We must show that if c and $a_1, a_2, \ldots, a_{k+1}$ are any real numbers, then $\prod_{i=1}^{k+1}(ca_i) = c^{k+1} \left(\prod_{i=1}^{k+1} a_i\right)$. Let c and $a_1, a_2, \ldots, a_{k+1}$ be any real numbers. Then

$\prod_{i=1}^{k+1}(c \cdot a_i)$

$\begin{aligned} &= \left(\prod_{i=1}^{k}(ca_i)\right)(ca_{k+1}) && \text{by the recursive definition of product} \\ &= \left(c^k \left(\prod_{i=1}^{k} a_i\right)\right)(ca_{k+1}) && \text{by substitution from the inductive hypothesis} \\ &= (c^k c)\left(\left(\prod_{i=1}^{k} a_i\right) a_{k+1}\right) && \text{by the associative and commutative laws of algebra} \\ &= c^{k+1} \left(\prod_{i=1}^{k+1} a_i\right) && \text{by the laws of exponents and the recursive} \\ & && \text{definition of product.} \end{aligned}$

[This is what was to be shown.]

24. *Proof (by mathematical induction)*: Let the property be the sentence "If A and $B_1, B_2, \ldots, B_n$ are any sets, then $A \cup \left(\bigcap_{i=1}^{n} B_i\right) = \bigcap_{i=1}^{n}(A \cup B_i)$."

Show that the property is true for $n = 1$: Let A and B_1 be any sets. By the recursive definition of intersection, both $A \cup \left(\bigcap_{i=1}^{1} B_i\right)$ and $\bigcap_{i=1}^{1}(A \cup B_i)$ equal $A \cup B_1$. Hence the property is true for $n = 1$.

Show that for all integers $k \geq 1$, if the property is true for $n = k$ then it is true for $n = k + 1$: Let k be an integer such that $k \geq 1$. Suppose that if A and $B_1, B_2, \ldots, B_k$ are any sets then $A \cup \left(\bigcap_{i=1}^{k} B_i\right) = \bigcap_{i=1}^{k}(A \cup B_i)$. *[This is the inductive hypothesis.]* We must show that if A and $B_1, B_2, \ldots, B_{k+1}$ are any sets then $A \cup \left(\bigcap_{i=1}^{k+1} B_i\right) = \bigcap_{i=1}^{k+1}(A \cup B_i)$. Let A and $B_1, B_2, \ldots, B_{k+1}$ be any sets. Then

$\begin{aligned} A \cup \left(\bigcap_{i=1}^{k+1} B_i\right) &= A \cup \left(\left(\bigcap_{i=1}^{k} B_i\right) \cap B_{k+1}\right) && \text{by the recursive definition of} \\ & && \text{intersection} \\ &= \left(A \cup \left(\bigcap_{i=1}^{k} B_i\right)\right) \bigcap (A \cup B_{k+1}) && \text{by the distributive laws for sets} \\ & && \text{(Theorem 5.2.2(3))} \\ &= \bigcap_{i=1}^{k}(A \cup B_i) \cap (A \cup B_{k+1}) && \text{by inductive hypothesis} \\ &= \bigcap_{i=1}^{k+1}(A \cup B_i) && \text{by the recursive definition of} \\ & && \text{intersection.} \end{aligned}$

[This is what was to be shown.]

27. *Proof 1 (by a variation of strong mathematical induction)*: Consider the property "$M(n) = 91$."

Show that the property is true for all integers n with $91 \leq n \leq 101$: This statement is proved above in the solution to exercise 26.

Show that for all integers k, if $1 \leq k < 91$ and the property is true for all i with $k < i \leq 101$, then it is true for k: Let k be an integer such that $1 \leq k < 91$ and suppose $M(i) = 91$ for all i with $k < i \leq 101$. *[This is the inductive hypothesis.]* Then $M(k) = M(M(k + 11))$ by definition of M and $12 \leq k + 11 < 102$. Thus $k < k + 11 \leq 101$, and so by inductive hypothesis $M(k + 11) = 91$. It follows that $M(k) = M(91)$, which equals 91 by the basis step above. Hence $M(91) = 91$ *[as was to be shown]*.

[Since the basis and inductive steps have been proved, it follows that $M(n) = 91$ for all integers $1 \leq n \leq 101$.]

Proof 2 (by contradiction): Suppose not. That is, suppose there is at least one positive integer $k \leq 101$ with $M(k) \neq 91$. Let q be the largest such integer. Then $1 \leq q \leq 101$ and $M(q) \neq 91$. Now by the solution to exercise 26, for each integer n with $91 \leq n \leq 101$, $M(n) = 91$. Thus $q \leq 90$ and so $q + 11 \leq 90 + 11 = 101$. Note, therefore, that $q + 11$ is a larger positive integer than q and is also less than 101. Hence, because q is the largest positive integer less than or equal to 101 with $M(q) \neq 91$, we must have that $M(q + 11) = 91$. But, by definition of M, $M(q) = M(M(q + 11))$. So $M(q) = M(M(q + 11)) = M(91) = 91$ (by the solution to exercise 26). Therefore, $M(q) \neq 91$ and $M(q) = 91$, which is a contradiction. We conclude that the supposition is false and $M(n) = 91$ for all positive integers $k \leq 101$.

30. (1) $T(2) = T(1) = 1$

   (2) $T(3) = T(10) = T(5) = T(16) = T(8) = T(4) = T(2) = 1$

   (3) $T(4) = 1$ by (2)

   (4) $T(5) = 1$ by (2)

   (5) $T(6) = T(3) = 1$ by (2)

   (6) $T(7) = T(22) = T(11) = T(34) = T(17) = T(52) = T(26) = T(13) = T(40) = T(20) = T(10) = 1$ by (2)

General Review for Chapter 8

Recursion

- What is an explicit formula for a sequence? *(p. 457)*
- What does it mean to define a sequence recursively? *(p. 457-8)*
- What is a recurrence relation with initial conditions? *(p. 458)*
- How do you compute terms of a recursively defined sequence? *(p. 458)*
- Can different sequences satisfy the same recurrence relation? *(p. 459)*
- What is the "recursive paradigm"? *(p. 460)*
- How do you develop recurrence relations for sequences that are variations of the towers of Hanoi sequence? *(p. 460)*
- How do you develop recurrence relations for sequences that are variations of the Fibonacci sequence? *(p. 464)*
- How do you develop recurrence relations for sequences that involve compound interest? *(p. 466-7)*
- How do you develop recurrence relations for sequences that involve the number of bit strings with a certain property? *(p. 467)*
- How do you find a recurrence relation for the number of ways a set of size n can be partitioned into r subsets? *(p. 469)*

Solving Recurrence Relations

- What is the method of iteration for solving a recurrence relation? *(p. 475)*
- How do you use the formula for the sum of the first n integers and the formula for the sum of the first n powers of a real number r to simplify the answers you obtain when you solve recurrence relations? *(p. 480)*
- How is mathematical induction used to check that the solution to a recurrence relation is correct? *(p. 483)*
- What is a second-order linear homogeneous recurrence relation with constant coefficients? *(p. 487)*
- What is the characteristic equation for a second-order linear homogeneous recurrence relation with constant coefficients? *(p. 489)*
- What is the distinct-roots theorem? If the characteristic equation of a relation has two distinct roots, how do you solve the relation? *(p. 491)*
- What is the single-root theorem? If the characteristic equation of a relation has a single root, how do you solve the relation? *(p. 497)*

General Recursive Definitions

- When a set is defined recursively, what are the three parts of the definition? *(p. 500)*
- Given a recursive definition for a set, how can you tell that a given element is in the set? *(p. 500-1)*
- What is structural induction? *(p. 502)*
- Given a recursive definition for a set, is there a way to tell that a given element is not in the set? *(p. 508, exercises 4, 6a, 8-14)*
- What are the recursive definitions for sum, product, union, and intersection? *(p. 503-5)*
- What is a recursive function? *(p. 505)*

Test Your Understanding: Chapter 8

Test yourself by filling in the blanks

1. The reason we can't always specify a sequence by giving its initial terms is that ____.

2. For a sequence $a_0, a_1, a_2, \ldots$ to be defined by an explicit formula means that ____.

3. A recurrence relation for a sequence $a_0, a_1, a_2, \ldots$ is ____. The initial conditions for such a recurrence relation specify ____.

4. To solve the Tower of Hanoi puzzle, we imagine the first step as consisting of all the moves needed to ____, the second step as moving ____, and the third step as ____.

5. The crucial observation used to solve the Fibonacci numbers problem is that the number of rabbits born at the end of month k is the same as ____. Also because no rabbits die, all the rabbits that are alive at the end of month $k - 1$ are still alive at ____. So the total number of rabbits alive at the end of month k equals ____.

6. When interest is compounded periodically, the amount in the account at the end of period k equals ____ plus ____.

7. A bit string of length k that does not contain the pattern 11 either starts with a ____, which is followed by a ____, or it starts with ____, which is followed by a ____.

8. The Stirling number of the second kind, $S_{n,r}$, can be interpreted as ____.

9. Because any partition of a set $X = \{x_1, x_2, \ldots, x_n\}$ either contains x_n or does not, the number of partitions of X into r subsets equals ____ plus ____.

10. To find an explicit formula for a recurrence relation by the method of iteration, you start by writing down ____ and then you use the recurrence relation to ____.

11. A sequence $a_1, a_2, a_3, \ldots$ is called an arithmetic sequence if, and only if, there is a constant d such that ____, or, equivalently, ____.

12. A sequence $a_1, a_2, a_3, \ldots$ is called an geometric sequence if, and only if, there is a constant r such that ____ or, equivalently, ____.

13. Two useful formulas for simplifying explicit formulas for recurrence relations that have been obtained by iteration are ____ and ____.

14. When an explicit formula for a recurrence relation has been obtained by iteration, the correctness of the formula can be checked by ____.

15. A second-order linear homogeneous recurrence relation with constant coefficients is a recurrence relation of the form ____ for all integers $k \geq$ ____, where ____.

16. Given a recurrence relation of the form $a_k = Aa_{k-1} + Ba_{k-2}$ for all integers $k \geq 2$, the characteristic equation of the relation is ____.

17. If a sequence $a_1, a_2, a_3, \ldots$ is defined by a second-order linear homogeneous recurrence relation with constant coefficients and the characteristic equation for the relation has two distinct roots r and s (which could be complex numbers), then the sequence satisfies an explicit formula of the form ____.

18. If a sequence $a_1, a_2, a_3, \ldots$ is defined by a second-order linear homogeneous recurrence relation with constant coefficients and the characteristic equation for the relation has only a single root r, then the sequence satisfies an explicit formula of the form ____.

19. The BASE for a recursive definition of a set is ____.

20. The RECURSION for a recursive definition of a set is ____.

21. The RESTRICTION for a recursive definition of a set is ____.

22. One way to show that a given element is in a recursively defined set is to start with an element or elements in the ____ and apply the rules from the ____ until you obtain the given element.

23. Another way to show that a given element is in a recursively defined set is to use ____ to characterize all the elements of the set and then observe that the given element satisfies the characterization.

24. To prove that every element in a recursively defined set S satisfies a certain property, you show that ____ and that, for each rule in the RECURSION, if____ then____.

25. A function is said to be defined recursively if, and only if, ____.

Answers

1. two sequences may have the same initial terms and yet have different terms later on

2. a_n equals an algebraic expression in the variable n

3. a formula that relates each term a_k of the sequence to certain of its predecessors; enough initial values of the sequence to enable subsequent values to be computed using the recurrence relation

4. transfer the top $k-1$ disks from the initial pole to the pole that is not the ultimate target pole;
   the bottom disk from the initial pole to the target pole;
   consisting of all the moves needed to transfer the top $k-1$ disks from the pole that is not the ultimate target pole to the target pole

5. the number of rabbits alive at the end of month $k-2$;
   the end of month k;
   the sum of the number of rabbits alive at the end of month $k-1$ plus the number alive at the end of month $k-2$

6. the amount in the account at the end of period $k-1$;
   the interest earned during period k

7. 0; bit string of length $k-1$ that does not contain the pattern 11;
   01; bit string of length $k-2$ that does not contain the pattern 11

8. the number of ways a set of size n can be partitioned into r subsets

9. the number of partitions of X into r subsets of which $\{x_n\}$ is one; the number of partitions of X into r subsets, none of which is $\{x_n\}$

10. as many terms of the sequence as are specified in the initial conditions; compute subsequent terms of the sequence by successive substitution

11. $a_k = a_{k-1} + d$, for all integers $k \geq 1$;
    $a_n = a_0 + dn$, for all integers $n \geq 0$

12. $a_k = ra_{k-1}$, for all integers $k \geq 1$;
    $a_n = a_0 r^n$, for all integers $n \geq 0$

13. the formula for the sum of the terms of a geometric sequence;
    the formula for the sum of the first n positive integers

14. mathematical induction

15. $a_k = Aa_{k-1} + Ba_{k-2}$; 2; A and B are fixed real numbers with $B \neq 0$

16. $t^2 - At - B = 0$

17. $a_n = Cr^n + Ds^n$, where C and D are real or complex numbers

18. $a_n = Cr^n + Dnr^n$, where C and D are real numbers

19. a statement that certain objects belong to the set

20. a collection of rules indicating how to form new set objects from those already known to be in the set

21. a statement that no objects belong to the set other than those coming from either the BASE or the RECURSION

22. BASE; RECURSION

23. structural induction

24. each object in the BASE satisfies the property
    the rule is applied to an object or objects in the BASE
    the object defined by the rule also satisfies the property

25. its rule of definition refers to itself

Chapter 9: The Efficiency of Algorithms

The focus of Chapter 9 is the analysis of algorithm efficiency in Sections 9.3 and 9.5. The chapter opens with a brief review of the properties of function graphs that are especially important for understanding O-, Ω-, and Θ-notations, which are introduced in Section 9.2. For simplicity, the examples in Section 9.2 are restricted to polynomial and rational functions. Section 9.3 introduces the analysis of algorithm efficiency with examples that include sequential search, insertion sort, selection sort (in the exercises), and polynomial evaluation (in the exercises). Section 9.4 discusses the properties of logarithms that are particularly important in the analysis of algorithms and other areas of computer science, and Section 9.5 applies the properties to analyze algorithms whose orders involve logarithmic functions. Examples in Section 9.5 include binary search and merge sort.

Section 9.1

3.

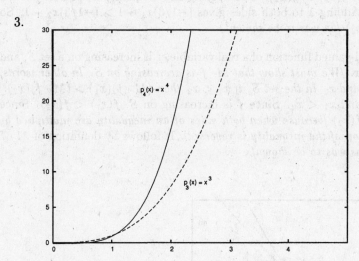

When $0 < x < 1$, $x^3 > x^4$. (For instance, $(1/2)^3 = 1/8 > 1/16 = (1/2)^4$.) When $x > 1$, $x^4 > x^3$.

6.

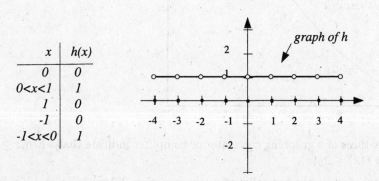

x	$h(x)$
0	0
$0<x<1$	1
1	0
-1	0
$-1<x<0$	1

12.

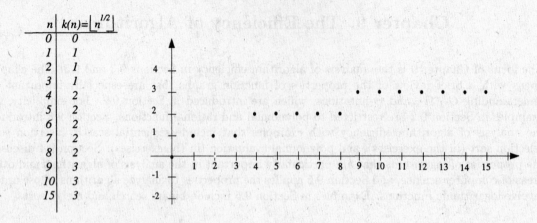

n	$k(n)=\lfloor n^{1/2}\rfloor$
0	0
1	1
2	1
3	1
4	2
5	2
6	2
7	2
8	2
9	3
10	3
15	3

15. *Proof*: Suppose that x_1 and x_2 are particular but arbitrarily chosen real numbers such that $x_1 < x_2$. *[We must show that $g(x_1) > g(x_2)$.]* Multiplying the inequality $x_1 < x_2$ by $-1/3$ gives $(-1/3)x_1 > (-1/3)x_2$. Adding 1 to both sides gives $(-1/3)x_1 + 1 > (-1/3)x_2 + 1$. So by definition of g, $g(x_1) > g(x_2)$ *[as was to be shown]*.

24. *Proof*: Suppose that f is a real-valued function of a real variable, f is increasing on a set S, and M is any negative real number. *[We must show that $M \cdot f$ is decreasing on S. In other words, we must show that for all x_1 and x_2 in the set S, if $x_1 < x_2$ then $(M \cdot f)(x_1) > (M \cdot f)(x_2)$.]* Suppose x_1 and x_2 are in S and $x_1 < x_2$. Since f is increasing on S, $f(x_1) < f(x_2)$. Since M is negative, $Mf(x_1) > Mf(x_2)$ *[because when both sides of an inequality are multiplied by a negative number, the direction of the inequality is reversed]*. It follows by definition of $M \cdot f$ that $(Mf)(x_1) > (Mf)(x_2)$ *[as was to be shown]*.

27.

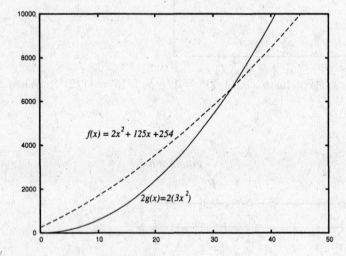

$f(x) = 2x^2 + 125x + 254$

$2g(x) = 2(3x^2)$

The zoom and trace features of a graphing calculator or computer indicate that when $x \geq 33.2$ (approximately), then $f(x) \leq 2g(x)$.

Alternatively, to find the answer algebraically, solve the equation $2(3x^2) = 2x^2 + 125x + 254$. Subtracting $2x^2 + 125x + 254$ from both sides gives $4x^2 - 125x - 254 = 0$. By the quadratic formula,

$$x = \frac{125 \pm \sqrt{125^2 + 4064}}{8} = \frac{125 \pm \sqrt{19689}}{8} \cong \frac{125 \pm 140.3175}{8},$$

and so $x \cong 33.165$ or $x \cong -1.915$. Since f and g are defined only for positive values of x, the only place where the two graphs cross is at $x \cong 33.165$. Thus if $x_0 = 33.2$, then for all $x > x_0$, $f(x) \leq 2g(x)$.

Section 9.2

3. *a. Formal version of negation*: $f(x)$ is not $\Theta(g(x))$ if, and only if, $\forall$ positive real numbers k, A, and B, $\exists$ a real number $x > k$ such that either $|f(x)| < A|g(x)|$ or $|f(x)| > B|g(x)|$.

   b. Informal version of negation: $f(x)$ is not $O(g(x))$ if, and only if, no matter what positive real numbers k, A, and B might be chosen, it is possible to find a real number x greater than k with the property that either $|f(x)| < A|g(x)|$ or $|f(x)| > B|g(x)|$.

9. Let $A = 1/2$, $B = 3$, and $k = 25$. Then by substitution, $A|x^2| \leq |3x^2 - 80x + 7| \leq B|x^2|$ for all $x > k$, and hence by definition of Θ-notation, $3x^2 - 80x + 7$ is $\Theta(x^2)$.

15. *a. Proof (by mathematical induction)*: Let the property $P(n)$ be the sentence "If x is any real number with $x > 1$, then $x^n > 1$.

    Show that the property is true for $n = 1$: We must show that if x is any real number with $x > 1$, then $x^1 > 1$. But $x > 1$ by hypothesis, and $x^1 = x$. So the property is true for $n = 1$.

    Show that for all integers $k \geq 1$, if the property is true for $n = k$ then it is true for $n = k + 1$: Let k be an integer with $k \geq 1$, and suppose that if x is any real number with $x > 1$, then $x^k > 1$. *[This is the inductive hypothesis.]* We must show that if x is any real number with $x > 1$, then $x^{k+1} > 1$. So suppose x is any real number with $x > 1$. By inductive hypothesis, $x^k > 1$, and multiplying both sides by the positive number x gives $x \cdot x^k > x \cdot 1$, or, equivalently, $x^{k+1} > x$. But $x > 1$, and so, by transitivity of order, $x^{k+1} > 1$ *[as was to be shown]*.

    b. Proof: Suppose x is any real number with $x > 1$ and m and n are integers with $m < n$. Then $n - m$ is an integer with $n - m \geq 1$, and so by part (a) $x^{n-m} > 1$. Multiplying both sides by x^m gives $x^m \cdot x^{n-m} > x^m \cdot 1$, and so, by the laws of exponents, $x^n > x^m$ *[as was to be shown]*.

21. *a.* For any real number $x > 1$,

$$\lfloor \lfloor \sqrt{x} \rfloor \rfloor \;=\; \lfloor \sqrt{x} \rfloor \quad \text{because since } x > 1 > 0, \text{ then } \lfloor \sqrt{x} \rfloor > 0$$
$$\Rightarrow \lfloor \lfloor \sqrt{x} \rfloor \rfloor \;\leq\; \sqrt{x} \quad \text{because } \lfloor r \rfloor \leq r \text{ for all real numbers } r$$
$$\Rightarrow \lfloor \lfloor \sqrt{x} \rfloor \rfloor \;\leq\; |\sqrt{x}| \quad \text{because } \sqrt{x} \geq 0.$$

    b. Suppose x is any real number with $x > 1$. By definition of floor, $\lfloor \sqrt{x} \rfloor \leq \sqrt{x} < \lfloor \sqrt{x} \rfloor + 1$. Now

$$\lfloor \sqrt{x} \rfloor + 1 \;\leq\; 2\lfloor \sqrt{x} \rfloor$$
$$\Leftrightarrow 1 \;\leq\; \lfloor \sqrt{x} \rfloor \quad \text{by subtracting } \lfloor \sqrt{x} \rfloor \text{ from both sides}$$
$$\Leftrightarrow 1 \;\leq\; \sqrt{x} \quad \text{by definition of floor}$$
$$\Leftrightarrow 1 \;\leq\; x \quad \text{by squaring both sides (okay because } x \text{ is positive),}$$

    and the last inequality is true because we are assuming that $1 < x$. Thus, $\sqrt{x} < \lfloor \sqrt{x} \rfloor + 1$ and $\lfloor \sqrt{x} \rfloor + 1 \leq 2\lfloor \sqrt{x} \rfloor$, and so, by the transitive law of order (Appendix A, T17), $\sqrt{x} \leq 2\lfloor \sqrt{x} \rfloor$. Dividing both sides by 2 gives $\frac{1}{2}\sqrt{x} \leq \lfloor \sqrt{x} \rfloor$. Finally, because all quantities are positive, we conclude that $\frac{1}{2}|\sqrt{x}| \leq |\lfloor \sqrt{x} \rfloor|$.

    c. Let $A = \frac{1}{2}$ and $a = 1$. Then by substitution, $A|\sqrt{x}| \leq |\lfloor \sqrt{x} \rfloor|$ for all $x > a$, and hence by definition of Ω-notation, $\lfloor \sqrt{x} \rfloor$ is $\Omega(\sqrt{x})$.

Let $B = 1$ and $b = 1$. Then $|\lfloor \sqrt{x} \rfloor| \leq B|\sqrt{x}|$ for all real numbers $x > b$, and so by definition of O-notation, $\lfloor \sqrt{x} \rfloor$ is $O(\sqrt{x})$.

d. By part (c) and Theorem 9.2.1(1), we can immediately conclude that $\lfloor \sqrt{x} \rfloor$ is $\Theta(\sqrt{x})$.

24. a. For all real numbers $x > 1$,

$$|\tfrac{1}{4}x^5 - 50x^3 + 3x + 12| \leq |\tfrac{1}{4}x^5| + |50x^3| + |3x| + |12| \qquad \text{by the triangle inequality}$$

$$\Rightarrow |\tfrac{1}{4}x^5 - 50x^3 + 3x + 12| \leq \tfrac{1}{4}x^5 + 50x^3 + 3x + 12 \qquad \begin{array}{l}\text{because } \tfrac{1}{4}x^5,\, 50x^3,\, 3x, \\ \text{and } 12 \text{ are positive}\end{array}$$

$$\Rightarrow |\tfrac{1}{4}x^5 - 50x^3 + 3x + 12| \leq \tfrac{1}{4}x^5 + 50x^5 + 3x^5 + 12x^5 \qquad \begin{array}{l}\text{because } x^3 < x^5,\, x < x^5, \\ \text{and } 1 < x^5 \text{for } x > 1\end{array}$$

$$\Rightarrow |\tfrac{1}{4}x^5 - 50x^3 + 3x + 12| \leq 66x^5 \qquad \text{because } \tfrac{1}{4} + 50 + 3 + 12 < 66$$

$$\Rightarrow |\tfrac{1}{4}x^5 - 50x^3 + 3x + 12| \leq 66|x^5| \qquad \text{because } x^5 \text{is positive.}$$

b. Let $B = 66$ and $b = 1$. Then by substitution, $|\tfrac{1}{4}x^5 - 50x^3 + 3x + 12| \leq B|x^2|$ for all $x > b$. Hence by definition of O-notation, $\tfrac{1}{4}x^5 - 50x^3 + 3x + 12$ is $O(x^2)$.

27. *Proof*: Suppose $a_0, a_1, a_2, \ldots, a_n$ are real numbers and $a_n \neq 0$; and let

$$d = 2 \left(\frac{|a_0| + |a_1| + |a_2| + \cdots + |a_{n-1}|}{|a_n|} \right).$$

Let a be greater than or equal to the maximum of d and 1. Then if $x > a$

$$x \geq 2 \left(\frac{|a_0| + |a_1| + |a_2| + \cdots + |a_3| + |a_{n-1}|}{|a_n|} \right)$$

$$\Rightarrow \quad \tfrac{1}{2}|a_n|x \geq |a_0| + |a_1| + |a_2| + \cdots + |a_{n-1}|$$

$$\text{by multiplying both sides by } \tfrac{1}{2}|a_n|$$

$$\Rightarrow \quad (1 - \tfrac{1}{2})|a_n|x \geq |a_0| \cdot \frac{1}{x^{n-1}} + |a_1| \cdot \frac{1}{x^{n-2}} + |a_2| \cdot \frac{1}{x^{n-3}} + \cdots + |a_{n-2}| \cdot \frac{1}{x} + |a_{n-1}| \cdot 1$$

$$\begin{array}{l}\text{because by exercise 15, when } x > 1 \text{ and } m \geq 1, \\ \text{then } x^m > 1, \text{ and so } 1 > \frac{1}{x^m}\end{array}$$

$$\Rightarrow \quad |a_n|x^n - \frac{|a_n|}{2}x^n \geq |a_0| + |a_1|x + |a_2|x^2 + \cdots + |a_{n-2}|x^{n-2} + |a_{n-1}|x^{n-1}$$

$$\text{by multiplying both sides by } x^{n-1}.$$

Subtracting all terms on the right-hand side from both sides and adding the second term on the left-hand side to both sides gives

$$|a_n|x^n - |a_{n-1}|x^{n-1} - |a_{n-2}|x^{n-2} - \cdots - |a_2|x^2 - |a_1|x - |a_0| \geq \frac{|a_n|}{2}x^n.$$

Now $-|r| \leq r \leq |r|$ for all real numbers r. Thus, when $x > 1$, $-|a_i|x^i \leq a_i x^i \leq |a_i|x^i$ for each integer i with $0 \leq i \leq n - 1$, and so

$$\begin{aligned}|a_n|x^n &- |a_{n-1}|x^{n-1} - |a_{n-2}|x^{n-2} - \cdots - |a_2|x^2 - |a_1|x - |a_0| \\ &\leq a_n x^n + a_{n-1}x^{n-1} + a_{n-2}x^{n-2} + \cdots + a_2 x^2 + a_1 x + a_0 \\ &\leq |a_n x^n + a_{n-1}x^{n-1} + a_{n-2}x^{n-2} + \cdots + a_2 x^2 + a_1 x + a_0|.\end{aligned}$$

Hence

$$\frac{|a_n|}{2}x^n \leq |a_n x^n + a_{n-1}x^{n-1} + a_{n-2}x^{n-2} + \cdots + a_2 x^2 + a_1 x + a_0|.$$

Let $A = \dfrac{|a_n|}{2}$ and let a be as defined above. Then

$$Ax^n \leq |a_n x^n + a_{n-1}x^{n-1} + a_{n-2}x^{n-2} + \cdots + a_2 x^2 + a_1 x + a_0| \quad \text{for all real numbers } x > a.$$

It follows by definition of Ω-notation that $a_n x^n + a_{n-1}x^{n-1} + a_{n-2}x^{n-2} + \cdots + a_2 x^2 + a_1 x + a_0$ is $\Omega(x^n)$.

30. Let $a = 2\left(\dfrac{50+3+12}{1/4}\right) = 520$, and let $A = \dfrac{1}{2} \cdot \dfrac{1}{4} = \dfrac{1}{8}$. If $x > 520$, then

$$x \ \geq \ 2\left(\frac{50+3+12}{1/4}\right)$$

$$\Rightarrow \qquad \frac{1}{2}\cdot\frac{1}{4}x \ \geq \ 50+3+12$$

by multiplying both sides by $\frac{1}{2}\cdot\frac{1}{4}$

$$\Rightarrow \qquad (1-\frac{1}{2})\cdot\frac{1}{4}x \ \geq \ 50\frac{1}{x}+3\cdot\frac{1}{x^3}+12\cdot\frac{1}{x^4}$$

because $1-\frac{1}{2}=\frac{1}{2}$, and, since $x > 520 > 1$, then $1 > \frac{1}{x}, 1 > \frac{1}{x^3}$ and $1 > \frac{1}{x^4}$

$$\Rightarrow \qquad \frac{1}{4}x^5 - \frac{1}{2\cdot}\frac{1}{4}x^5 \ \geq \ 50x^3 + 3x + 12$$

by multiplying both sides by x^4

$$\Rightarrow \qquad \frac{1}{4}x^5 - 50x^3 - 3x - 12 \ \geq \ \frac{1}{2}\cdot\frac{1}{4}x^5$$

by subtracting $50x^3 + 3x + 12$ from and adding $\frac{1}{2}\cdot\frac{1}{4}x^5$ to both sides

$$\Rightarrow \qquad \frac{1}{4}x^5 - 50x^3 + 3x + 12 \ \geq \ \frac{1}{2}\cdot\frac{1}{4}x^5$$

because $3x + 12 > -3x - 12$ since $x > 0$

$$\Rightarrow \qquad \left|\frac{1}{4}x^5 - 50x^3 + 3x + 12\right| \ \geq \ \frac{1}{8}\left|x^5\right|$$

because both sides are nonnegative.

Thus for all real numbers $x > a$, $\left|\dfrac{1}{4}x^5 - 50x^3 + 3x + 12\right| \geq A\left|x^5\right|$. Hence, by definition of Ω-notation, we conclude that $\dfrac{1}{4}x^5 - 50x^3 + 3x + 12$ is $\Omega(x^5)$.

33. By exercise 24, $\frac{1}{4}x^5 - 50x^3 + 3x + 12$ is $O(x^5)$ and by exercise 31 $\frac{1}{4}x^5 - 50x^3 + 3x + 12$ is $\Omega(x^5)$. Thus, by Theorem 9.2.1(1), $\frac{1}{4}x^5 - 50x^3 + 3x + 12$ is $\Theta(x^5)$.

36. $\dfrac{x(x-1)}{2} + 3x = \dfrac{x^2-x}{2}+\dfrac{6}{2}x = \dfrac{1}{2}x^2 + \dfrac{5}{2}x$ is $\Theta(x^2)$ by the theorem on polynomial orders.

39. $2(n-1)+\dfrac{n(n+1)}{2} + 4\left(\dfrac{n(n-1)}{2}\right) = 2n-2+\dfrac{n^2}{2}+\dfrac{n}{2}+2(n^2-n) = \dfrac{5}{2}n^2+\dfrac{1}{2}n-2$ is $\Theta(n^2)$ by the theorem on polynomial orders.

45. $\displaystyle\sum_{k=1}^{n}(k+3) = \sum_{k=1}^{n} k + \sum_{k=1}^{n} 3$ *[by Theorem 4.1.1]* $= \dfrac{n(n+1)}{2} + \underbrace{(3+3+\cdots+3)}_{n \text{ terms}}$ *[by Theorem 4.2.2]*
$= \frac{1}{2}n^2 + \frac{1}{2}n + 3n$ *[because multiplication is repeated addition]* $= \frac{1}{2}n^2 + \frac{7}{2}n$, which is $\Theta(n^2)$ by the theorem on polynomial orders.

48. *a.Proof*: Suppose $a_0, a_1, a_2, \ldots, a_n$ are real numbers and $a_n \neq 0$. Then

$$\lim_{x\to\infty}\left|\frac{a_n x^n + a_{n-1}x^{n-1} + \cdots + a_2 x^2 + a_1 x + a_0}{a_n x^n}\right|$$

$$= \lim_{x\to\infty}\left(1 + \left|\frac{a_{n-1}}{a_n}\right|\frac{1}{x} + \cdots + \left|\frac{a_2}{a_n}\right|\frac{1}{x^{n-2}} + \left|\frac{a_1}{a_n}\right|\frac{1}{x^{n-1}} + \left|\frac{a_0}{a_n}\right|\frac{1}{x^n}\right)$$

$$= \lim_{x\to\infty}1 + \left|\frac{a_{n-1}}{a_n}\right|\lim_{x\to\infty}\frac{1}{x} + \cdots + \left|\frac{a_2}{a_n}\right|\lim_{x\to\infty}\frac{1}{x^{n-2}} + \left|\frac{a_1}{a_n}\right|\lim_{x\to\infty}\frac{1}{x^{n-1}} + \left|\frac{a_0}{a_n}\right|\lim_{x\to\infty}\frac{1}{x^n}$$

$$= 1$$

because $\lim_{x \to \infty} \dfrac{1}{x^k} = 0$ for all integers $k \geq 1$.

b. Proof: Suppose $a_0, a_1, a_2, \ldots, a_n$ are real numbers and $a_n \neq 0$. By part (a) and the definition of limit, we can make the following statement: For all positive real numbers ε, there exists a real number M (which we may take to be positive) such that

$$1 - \varepsilon < \left| \frac{a_n x^n + a_{n-1} x^{n-1} + \cdots + a_2 x^2 + a_1 x + a_0}{a_n x^n} \right| < 1 + \varepsilon \qquad \text{for all real numbers } x > M.$$

Let $\varepsilon = 1/2$. Then there exists a real number M_0 such that

$$1 - \frac{1}{2} < \left| \frac{a_n x^n + a_{n-1} x^{n-1} + \cdots + a_2 x^2 + a_1 x + a_0}{a_n x^n} \right| < 1 + \frac{1}{2} \qquad \text{for all real numbers } x > M_0.$$

Equivalently, for all real numbers $x > M_0$,

$$\frac{1}{2} |a_n| |x^n| < \left| a_n x^n + a_{n-1} x^{n-1} + \cdots + a_2 x^2 + a_1 x + a_0 \right| < \frac{3}{2} |a_n| |x^n|.$$

Let $A = \frac{1}{2} |a_n|$, $B = \frac{3}{2} |a_n|$, and $k = M_0$. Then

$$A |x^n| \leq \left| a_n x^n + a_{n-1} x^{n-1} + \cdots + a_2 x^2 + a_1 x + a_0 \right| \leq B |x^n| \qquad \text{for all real numbers } x > k,$$

and so, by definition of Θ-notation, $a_n x^n + a_{n-1} x^{n-1} + \cdots + a_2 x^2 + a_1 x + a_0$ is $\Theta(x^n)$.

57. *b. Proof (by mathematical induction):* Let the property $P(n)$ be the inequality

$$\frac{1}{2} n^{3/2} \leq \sqrt{1} + \sqrt{2} + \sqrt{3} + \cdots + \sqrt{n}.$$

Show that the property is true for $n = 1$: We must show that $\frac{1}{2} \cdot 1^{3/2} \leq \sqrt{1}$. But the left-hand side of the inequality is $1/2$ and the right-hand side is 1, and $1/2 < 1$. So the property is true for $n = 1$.

Show that for all integers $k \geq 1$, if the property is true for $n = k$ then it is true for $n = k + 1$: Let k be an integer with $k \geq 1$, and suppose that

$$\frac{1}{2} k^{3/2} \leq \sqrt{1} + \sqrt{2} + \sqrt{3} + \cdots + \sqrt{k}. \qquad \textit{[This is the inductive hypothesis.]}$$

We must show that

$$\frac{1}{2} (k + 1)^{3/2} \leq \sqrt{1} + \sqrt{2} + \sqrt{3} + \cdots + \sqrt{k + 1}.$$

By adding $\sqrt{k + 1}$ to both sides of the inductive hypothesis, we have

$$\frac{1}{2} k^{3/2} + \sqrt{k + 1} \leq \sqrt{1} + \sqrt{2} + \sqrt{3} + \cdots + \sqrt{k} + \sqrt{k + 1}.$$

Thus, by the transitivity of order, it suffices to show that

$$\frac{1}{2} (k + 1)^{3/2} \leq \frac{1}{2} k^{3/2} + \sqrt{k + 1}.$$

Now when $k \geq 1$, $k^2 \geq k^2 - 1 = (k - 1)(k + 1)$. Divide both sides by $k(k - 1)$ to obtain $\dfrac{k}{k - 1} \geq \dfrac{k + 1}{k}$. But $\dfrac{k + 1}{k} \geq 1$, and any number greater than or equal to 1 is greater than or equal to its own square root. Thus $\dfrac{k}{k - 1} \geq \dfrac{k + 1}{k} \geq \sqrt{\dfrac{k + 1}{k}} = \dfrac{\sqrt{k + 1}}{\sqrt{k}}$. Hence $k\sqrt{k} \geq (k - 1)\sqrt{k + 1} = (k + 1 - 2)\sqrt{k + 1} = (k + 1)^{3/2} - 2\sqrt{k + 1}$. Multiplying both sides by $1/2$ gives $\frac{1}{2} k^{3/2} \geq \frac{1}{2}(k + 1)^{3/2} - \sqrt{k + 1}$, or, equivalently, $\frac{1}{2}(k + 1)^{3/2} \leq \frac{1}{2} k^{3/2} + \sqrt{k + 1}$. *[This is what was to be shown].*

60. *Proof*: Suppose $f(x)$ and $g(x)$ are $o(h(x))$ and a and b are any real numbers. Then by properties of limits,

$$\lim_{x \to \infty} \frac{af(x) + bg(x)}{h(x)} = a \lim_{x \to \infty} \frac{f(x)}{h(x)} + b \lim_{x \to \infty} \frac{g(x)}{h(x)} = a \cdot 0 + b \cdot 0 = 0.$$

So $af(x) + bg(x)$ is $o(h(x))$.

Section 9.3

3. *a.* When the input size is increased from m to $2m$, the number of operations increases from cm^3 to $c(2m)^3 = 8cm^3$.

   b. By part (a), the number of operations increases by a factor of $8cm^3/overcm^3 = 8$.

   c. When the input size is increased by a factor of 10 (from m to $10m$), the number of operations increases by a factor of $\dfrac{c(10m^3)}{cm^3} = \dfrac{1000cm^3}{cm^3} = 1000$.

12. *a.* For each iteration of the inner loop there is one comparison. The number of iterations of the inner loop can be deduced from the following table, which shows the values of k and i for which the inner loop is executed.

k	1				2				$\cdots$	$n-2$		$n-1$	
i	2	3	$\cdots$	n	3	4	$\cdots$	n	$\cdots$	$n-1$	n	n	

$$\underbrace{}_{n-1} \quad \underbrace{}_{n-2} \quad \underbrace{}_{2} \quad \underbrace{}_{1}$$

Therefore, by Theorem 4.2.2, the number of iterations of the inner loop is $(n-1) + (n-2) + \cdots + 2 + 1 = \frac{n(n-1)}{2}$. It follows that the total number of elementary operations that must be performed when the algorithm is executed is $1 \cdot \left(\frac{n(n-1)}{2}\right) = \frac{1}{2}n^2 - \frac{1}{2}n$.

By the theorem on polynomial orders, $\frac{1}{2}n^2 - \frac{1}{2}n$ is $\Theta(n^2)$, and so the algorithm segment has order n^2.

15. *a.* There are three multiplications for each iteration of the inner loop, and there is one additional addition for each iteration of the outer loop. The number of iterations of the inner loop can be deduced from the following table, which shows the values of i and j for which the inner loop is executed.

i	1				2				$\cdots$	$n-2$		$n-1$	n
j	2	3	$\cdots$	n	3	4	$\cdots$	n	$\cdots$	$n-1$	n	n	

$$\underbrace{}_{n-1} \quad \underbrace{}_{n-2} \quad \underbrace{}_{2} \quad \underbrace{}_{1}$$

Hence, by Theorem 4.2.2, the total number of iterations of the inner loop is $(n-1) + (n-2) + \cdots + 2 + 1 = \dfrac{n(n-1)}{2}$. Because three multiplications are performed for each iteration of the inner loop, the number of operations that are performed when the inner loop is executed is $3 \cdot \dfrac{n(n-1)}{2} = \frac{3}{2}(n^2 - n) = \frac{3}{2}n^2 - \frac{3}{2}n$. Now an additional operation is performed each time the outer loop is executed, and because the outer loop is executed n times, this gives an additional n operations. Therefore, the total number of operations is $(\frac{3}{2}n^2 - \frac{3}{2}n) + n = \frac{3}{2}n^2 - \frac{1}{2}n$.

   b. By the theorem on polynomial orders, $\frac{3}{2}n^2 - \frac{1}{2}n$ is $\Theta(n^2)$, and so the algorithm segment has order n^2.

18. *a.* There are n iterations of the inner loop for each iteration of the middle loop; there are $2n$ iterations of the middle loop for each iteration of the outer loop; and there are n iterations of the outer loop. Therefore, by the multiplication rule, there are $n \cdot 2n \cdot n = 2n^3$ iterations of the

inner loop. Because there are two multiplications for each iteration of the inner loop. the total number of elementary operations that must be performed when the algorithm is executed is $2 \cdot (2n^3) = 4n^3$.

b. By the theorem on polynomial orders, $4n^3$ is $\Theta(n^3)$, and so the algorithm segment has order n^3.

21.

	$a[1]$	$a[2]$	$a[3]$	$a[4]$	$a[5]$
initial order	7	3	6	9	5
result of step 1	3	7	6	9	5
result of step 2	3	6	7	9	5
result of step 3	3	6	7	9	5
result of step 4	3	5	6	7	9

27. a.

$$
\begin{aligned}
E_1 &= 0 \\
E_2 &= E_1 + 2 + 1 = 3 \\
E_3 &= E_2 + 3 + 1 = 3 + 4 \\
E_4 &= E_3 + 4 + 1 = 3 + 4 + 5 \\
E_5 &= E_4 + 5 + 1 = 3 + 4 + 5 + 6
\end{aligned}
$$

$$
\begin{aligned}
\text{Guess:} \quad E_n &= 3 + 4 + 5 + \cdots + (n+1) = [1 + 2 + 3 + 4 + 5 + \cdots + (n+1)] - (1+2) \\
&= \frac{(n+1)(n+2)}{2} - 3 = \frac{n^2 + 3n + 2 - 6}{2} = \frac{n^2 + 3n - 4}{2}
\end{aligned}
$$

b. *Proof by mathematical induction*: Let $E_1, E_2, E_3, \ldots$ be a sequence that satisfies the recurrence relation $E_k = E_{k-1} + k + 1$ for all integers $k \geq 2$, with initial condition $E_1 = 0$, and let the property $P(n)$ be the equation $E_n = \dfrac{n^2 + 3n - 4}{2}$.

Show that the property is true for $n = 1$: For $n = 1$ the equation is $E_1 = \dfrac{1^2 + 3 \cdot 1 - 4}{2} = 0$, which is true.

Show that for all integers $k \geq 1$, if the property is true for $n = k$ then it is true for $n = k + 1$: Let k be an integer with $k \geq 1$, and suppose that $E_k = \dfrac{k^2 + 3k - 4}{2}$. *[This is the inductive hypothesis.]* We must show that $E_{k+1} = \dfrac{(k+1)^2 + 3(k+1) - 4}{2}$. But the left-hand side of this equation equals

$$
\begin{aligned}
E_{k+1} &= E_k + (k+1) + 1 && \text{by definition of } E_1, E_2, E_3, \ldots \\
&= \frac{k^2 + 3k - 4}{2} + (k+1) + 1 && \text{by inductive hypothesis} \\
&= \frac{k^2 + 3k - 4 + 2k + 4}{2} \\
&= \frac{k^2 + 5k}{2}.
\end{aligned}
$$

And the right-hand side of the equation equals $\dfrac{(k+1)^2 + 3(k+1) - 4}{2} = \dfrac{k^2 + 2k + 1 + 3k + 3 - 4}{2} = \dfrac{k^2 + 5k}{2}$ also.

[This is what was to be shown.]

33. There is one comparison for each combination of values of k and i in the trace table for exercise 31. This gives a total of 10.

39. By the result of exercise 38, $s_n = \frac{1}{2}n^2 + \frac{3}{2}n$, which is $\Theta(n^2)$ by the theorem on polynomial orders.

42. There are two operations (one addition and one multiplication) per iteration of the loop, and there are n iterations of the loop. Therefore, $t_n = 2n$.

Section 9.4

6.

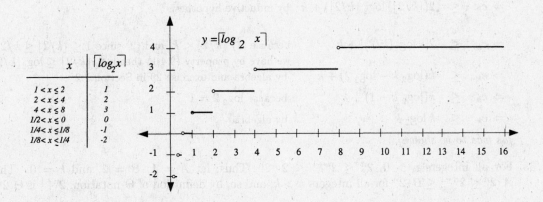

x	$\lceil \log_2 x \rceil$
$1 < x \leq 2$	1
$2 < x \leq 4$	2
$4 < x \leq 8$	3
$1/2 < x \leq 0$	0
$1/4 < x \leq 1/8$	-1
$1/8 < x \leq 1/4$	-2

12. When $\frac{1}{2} < x < 1$, then $-1 < \log_2 x < 0$.

When $\frac{1}{4} < x < \frac{1}{2}$, then $-2 < \log_2 x < -1$.

When $\frac{1}{8} < x < \frac{1}{4}$, then $-3 < \log_2 x < -2$.

And so forth.

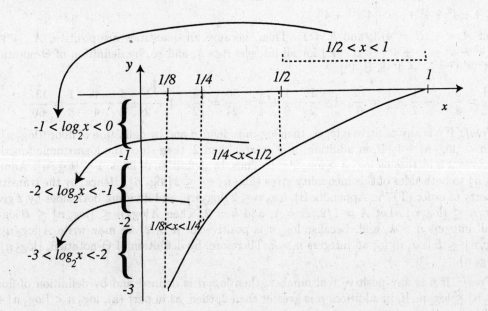

24. *Proof (by strong mathematical induction):* Let $c_1, c_2, c_3, \ldots$ be a sequence that satisfies the recurrence relation $c_k = 2c_{\lfloor k/2 \rfloor} + k$ for all integers $k \geq 2$, with initial condition $c_1 = 0$, and let the property $P(n)$ be the inequality $c_n \leq n \log_2 n$.

Show that the property is true for $n = 1$: For $n = 1$ the inequality states that $c_1 \leq 1 \cdot \log_2 1 = 1 \cdot 0 = 0$, which is true because $c_1 = 0$.

Show that for all integers $k \geq 1$, if the property is true for all i with $1 \leq i < k$ then it is true for k: Let k be an integer with $k > 1$ and suppose $c_i \leq i \log_2 i$ for all integers i with $1 \leq i < k$. *[This is the inductive hypothesis.]* We must show that $c_k \leq k \log_2 k$. First note that because k is an integer with $k > 1$, $2 \leq k$. Thus $1 \leq \dfrac{k}{2} < \dfrac{k}{2} + \dfrac{k}{2} = k$, and so

$1 \leq \left\lfloor \dfrac{k}{2} \right\rfloor < k$. Also note that $\lfloor k/2 \rfloor \leq k/2$ by definition of floor. Then

$$
\begin{aligned}
c_k &= 2c_{\lfloor k/2 \rfloor} + k && \text{by definition of } c_1, c_2, c_3, \ldots \\
\Rightarrow c_k &\leq 2(\lfloor k/2 \rfloor) \log_2 \lfloor k/2 \rfloor) + k && \text{by inductive hypothesis} \\
\Rightarrow c_k &\leq 2[k \log_2(k/2)] + k && \text{because (1) } \lfloor k/2 \rfloor < k, \text{ and (2) since } 1 \leq \lfloor k/2 \rfloor \leq k/2, \\
&&& \text{we have by property (9.4.1) that } \log_2 \lfloor k/2 \rfloor \leq \log_2(k/2) \\
\Rightarrow c_k &\leq k(\log_2 k - \log_2 2) + k && \text{by algebra and exercise 29 in Section 7.2} \\
\Rightarrow c_k &\leq k(\log_2 k - 1) + k && \text{because } \log_2 2 = 1 \\
\Rightarrow c_k &\leq k \log_2 k && \text{by algebra,}
\end{aligned}
$$

[as was to be shown].

33. For all integers $n > 0$, $2^n \leq 2^{n+1} \leq 2 \cdot 2^n$. Thus, let $A = 1$, $B = 2$, and $k = 0$. Then $A \cdot 2^n \leq 2^{n+1} \leq B \cdot 2^n$ for all integers $n > k$, and so, by definition of Θ-notation, 2^{n+1} is $\Theta(2^n)$.

36. By factoring out a 4 and using the formula for the sum of a geometric sequence (Theorem 4.2.3), we have that for all integers $n > 1$,

$$4 + 4^2 + 4^3 + \cdots + 4^n = 4(1 + 4 + 4^2 + \cdots + 4^{n-1})$$

$$= 4\left(\frac{4^{(n-1)+1} - 1}{4 - 1}\right) = \frac{4}{3}(4^n - 1) = \frac{4}{3} \cdot 4^n - \frac{4}{3} \leq \frac{4}{3} \cdot 4^n.$$

Moreover, because $4 + 4^2 + 4^3 + \cdots + 4^{n-1} \geq 0$,

$$4^n \leq 4 + 4^2 + 4^3 + \cdots + 4^{n-1} + 4^n.$$

So let $A = 1$, $B = 4/3$, and $k = 1$. Then, because all quantities are positive, $A \cdot |4^n| \leq |4 + 4^2 + 4^3 + \cdots + 4^n| \leq B \cdot |4^n|$ for all integers $n > k$, and so, by definition of Θ-notation, $4 + 4^2 + 4^3 + \cdots + 4^n$ is $\Theta(4^n)$.

42. $1 + \dfrac{1}{2} = \dfrac{3}{2}$, $1 + \dfrac{1}{2} + \dfrac{1}{3} = \dfrac{11}{6}$, $1 + \dfrac{1}{2} + \dfrac{1}{3} + \dfrac{1}{4} = \dfrac{50}{24} = \dfrac{25}{12}$, $1 + \dfrac{1}{2} + \dfrac{1}{3} + \dfrac{1}{4} + \dfrac{1}{5} = \dfrac{137}{60}$

45. *a. Proof:* If n is any positive integer, then $\log_2 n$ is defined and by definition of floor, $\lfloor \log_2 n \rfloor \leq \log_2 n < \lfloor \log_2 n \rfloor + 1$. If, in addition, n is greater than 2, then since the logarithmic function with base 2 is increasing $\log_2 n > \log_2 2 = 1$. Thus, by definition of floor, $1 \leq \lfloor \log_2 n \rfloor$. Adding $\lfloor \log_2 n \rfloor$ to both sides of this inequality gives $\lfloor \log_2 n \rfloor + 1 \leq 2 \lfloor \log_2 n \rfloor$. Hence, by the transitive property of order (T17 in Appendix B), $\log_2 n \leq 2 \lfloor \log_2 n \rfloor$, and dividing both sides by 2 gives $\frac{1}{2} \log_2 n \leq \lfloor \log_2 n \rfloor$. Let $A = 1/2$, $B = 1$, and $k = 2$. Then $A \log_2 n \leq \lfloor \log_2 n \rfloor \leq B \log_2 n$ for all integers $n > k$, and, because $\log_2 n$ is positive for $n > 2$, we may write $A |\log_2 n| \leq |\lfloor \log_2 n \rfloor| \leq B |\log_2 n|$ for all integers $n > k$. Therefore, by definition of Θ-notation, $\lfloor \log_2 n \rfloor$ is $\Theta(\log_2 n)$.

b. Proof: If n is any positive real number, then $\log_2 n$ is defined and by definition of floor, $\lfloor \log_2 n \rfloor \leq \log_2 n$. If, in addition, n is greater than 2, then, as in part (a), $\log_2 n < \lfloor \log_2 n \rfloor + 1$

and $\lfloor \log_2 n \rfloor + 1 \leq 2\log_2 n$. Hence, because $\log_2 n$ is positive for $n > 2$, we may write $|\log_2 n| \leq |\lfloor \log_2 n \rfloor + 1| \leq 2|\log_2 n|$. Let $A = 1$, $B = 2$ and $k = 2$. Then $A|\log_2 n| \leq |\lfloor \log_2 n \rfloor + 1| \leq B|\log_2 n|$ for all integers $n > k$. Therefore, by definition of Θ-notation, $\lfloor \log_2 n \rfloor + 1$ is $\Theta(\log_2 n)$.

48. *Proof*: Suppose n is a variable that takes positive integer values. Then whenever $n \geq 2$,

$$2^n = \underbrace{2 \cdot 2 \cdot 2 \cdot 2 \cdot 2 \cdots 2}_{n \text{ factors}} \leq \underbrace{2 \cdot 2 \cdot 3 \cdot 4 \cdot 5 \cdots n}_{n \text{ factors}} \leq 2n!.$$

Let $B = 2$ and $b = 2$. Since 2^n and $n!$ are positive for all n, $|2^n| \leq B|n!|$ for all integers $n > b$. Hence by definition of O-notation 2^n is $O(n!)$.

51. *a.* Let n be any positive integer. Then for any real number x *[because $u < 2^u$ for all real numbers u]*,

$$x/n < 2^{x/n} \implies x < n2^{x/n} \implies x^n < (n2^{x/n})^n = n^n \cdot 2^x.$$

So $x^n < n^n 2^x$.

b. Let x be any positive real number and let n be any positive integer. Then $x^n = |x^n|$ and $n^n 2^x = 2^x |n^n|$, and thus the result of part (a) may be written as $|x^n| \leq 2^x |n^n|$. Let $B = 2^x$ and $b = 0$. Then $|x^n| \leq B|n^n|$ for all integers $n > b$, and so by definition of O-notation x^n is $O(n^n)$.

Section 9.5

6. *a.*

index	0			
bot	1	1	1	1
top	10	4	1	0
mid		5	2	1

b.

index	0		8
bot	1	6	
top	10		
mid		5	8

12.

n	424	141	47	15	5	1	0

15. If $n \geq 3$, then, by definition of floor, $\lfloor \log_3 n \rfloor \leq \log_3 n < \lfloor \log_3 n \rfloor + 1$. Thus $b_n = 1 + \lfloor \log_3 n \rfloor \leq 1 + \log_3 n$ *[by definition of floor]* $\leq \log_3 n + \log_3 n$ *[because if $n \geq 3$ then $\log_3 n \geq 1$]* $= 2\log_3 n$. Furthermore, because $\log_3 n \geq 0$ for $n > 2$, we may write $|\log_3 n| < |\lfloor \log_3 n \rfloor + 1| \leq 2|\log_3 n|$. Let $A = 1$, $B = 2$, and $k = 2$. Then $A|\log_3 n| < |\lfloor \log_3 n \rfloor + 1| \leq B|\log_3 n|$ for all integers $n > k$. Hence by definition of Θ-notation, $b_n = 1 + \lfloor \log_3 n \rfloor$ is $\Theta(\log_3 n)$, and thus the algorithm segment has order $\log_3 n$.

18. Suppose an array of length k is input to the **while** loop and the loop is iterated one time. The elements of the array can be matched with the integers from 1 to k with $m = \left\lceil \dfrac{k+1}{2} \right\rceil$, as shown below:

	left subarray				right subarray		
$a[bot]$	$a[bot+1]$ $\cdots$	$a[mid-1]$	$a[mid]$	$a[mid+1]$ $\cdots$	$a[top-1]$	$a[top]$
$\updownarrow$	$\updownarrow$	$\updownarrow$	$\updownarrow$	$\updownarrow$	$\updownarrow$	$\updownarrow$
1	2	$m-1$	m	$m+1$	$k-1$	k

Case 1 (k is even): In this case $m = \left\lceil \dfrac{k+1}{2} \right\rceil = \left\lceil \dfrac{k}{2} + \dfrac{1}{2} \right\rceil = \dfrac{k}{2} + 1$, and so the number of elements in the left subarray equals $m - 1 = (\dfrac{k}{2} + 1) - 1 = \dfrac{k}{2} = \left\lfloor \dfrac{k}{2} \right\rfloor$. The number of elements in the right subarray equals $k - (m+1) - 1 = k - m = k - (\dfrac{k}{2} + 1) = \dfrac{k}{2} - 1 < \left\lfloor \dfrac{k}{2} \right\rfloor$. Hence both subarrays (and thus the new input array) have length at most $\left\lfloor \dfrac{k}{2} \right\rfloor$.

Case 2 (k is odd): In this case $m = \left\lceil \dfrac{k+1}{2} \right\rceil = \dfrac{k+1}{2}$, and so the number of elements in the left subarray equals $m - 1 = \dfrac{k+1}{2} - 1 = \dfrac{k-1}{2} = \left\lfloor \dfrac{k}{2} \right\rfloor$. The number of elements in the right subarray equals $k - m = k - \dfrac{k+1}{2} = \dfrac{k-1}{2} = \left\lfloor \dfrac{k}{2} \right\rfloor$ also. Hence both subarrays (and thus the new input array) have length $\left\lfloor \dfrac{k}{2} \right\rfloor$.

The arguments in cases 1 and 2 show that the length of the new input array to the next iteration of the **while** loop has length at most $\lfloor k/2 \rfloor$.

21.

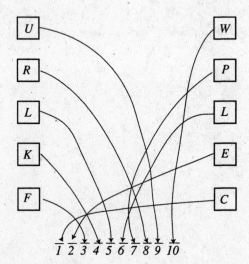

24. *a.* Refer to Figure 9.5.3. Observe that when k is odd, the subarray $a[mid+1], a[mid+2], \ldots a[top]$ has length $k - \left(\dfrac{k+1}{2} + 1 \right) + 1 = \dfrac{k-1}{2} = \lfloor k/2 \rfloor$. And when k is even, the subarray $a[mid+1], a[mid+2], \ldots a[top]$ has length $k - \left(\dfrac{k}{2} + 1 \right) + 1 = \dfrac{k}{2} = \lfloor k/2 \rfloor$. So in either case the subarray has length $\lfloor k/2 \rfloor$.

General Review for Chapter 9

Definitions: How are the following terms defined?

- real-valued function of a real variable *(p. 510)*
- graph of a real-valued function of a real variable *(p. 511)*
- power function with exponent a *(p. 511)*
- floor function *(p. 512)*
- multiple of a real-valued function of a real variable *(p. 514)*
- increasing function *(pp. 515-6)*
- decreasing function *(pp. 515-6)*
- $f(x)$ is $\Omega(g(x))$, where f and g are real-valued functions of a real variable defined on the same set of nonnegative real numbers *(p. 519)*
- $f(x)$ is $O(g(x))$, where f and g are real-valued functions of a real variable defined on the same set of nonnegative real numbers *(p. 519)*
- $f(x)$ is $\Theta(g(x))$, where f and g are real-valued functions of a real variable defined on the same set of nonnegative real numbers *(p. 519)*
- algorithm A is $\Theta(g(n))$ (or A has order $g(n)$) *(p. 533)*
- algorithm A is $\Omega(g(n))$ (or A has a best case order $g(n)$) *(p. 533)*
- algorithm A is $O(g(n))$ (or A has a worst case order $g(n)$) *(p. 533)*

Polynomial and Rational Functions and Their Orders

- What is the difference between the graph of a function defined on an interval of real numbers and the graph of a function defined on a set of integers? *(p. 513)*
- How do you graph a multiple of a real-valued function of a real variable? *(p. 514)*
- How do you prove that a function is increasing (decreasing)? *(p. 516)*
- What are some properties of O-, Ω-, and Θ-notation? Can you prove them? *(p. 521)*
- If $x > 1$, what is the relationship between x^r and x^s, where r and s are rational numbers and $r < s$? *(p. 522)*
- Given a polynomial, how do you use the definition of Θ-notation to show that the polynomial has order x^n, where n is the degree of the polynomial? *(pp. 523-5)*
- What is the theorem on polynomial orders? *(p. 526)*
- What is an order for the sum of the first n integers? *(p. 527)*

Efficiency of Algorithms

- How do you compute the order of an algorithm segment that contains a loop? a nested loop? *(pp. 533-35)*
- How do you find the number of times a loop will iterate when an algorithm segment is executed? *(p. 534)*
- How do you use the theorem on polynomial orders to help find the order of an algorithm segment? *(p. 535)*
- What is the sequential search algorithm? How do you compute its worst case order? its average case order? *(p. 536)*
- What is the insertion sort algorithm? How do you compute its best and worst case orders? *(p. 536)*

Logarithmic and Exponential Orders

- What do the graphs of logarithmic and exponential functions look like? *(pp. 544-5)*

- What can you say about the base 2 logarithm of a number that is between two consecutive powers of 2? *(p. 546)*
- How do you compute the number of bits needed to represent a positive integer in binary notation? *(p. 547)*
- How are logarithms used to solve recurrence relations? *(p. 548)*
- If $b > 1$, what can you say about the relation among $\log_b x$, x^r, and $x \log_b x$? *(p. 550)*
- If $b > 1$ and $c > 1$, how are orders of $\log_b x$ and $\log_c x$ related? *(p. 552)*
- What is an order for a harmonic sum? *(p. 553)*
- What is a divide-and-conquer algorithm? *(p. 557)*
- What is the binary search algorithm? *(p. 557)*
- What is the worst case order for the binary search algorithm, and how do you find it? *(p. 560)*
- What is the merge sort algorithm? *(p. 564)*
- What is the worst case order for the merge sort algorithm, and how do you find it? *(p. 567)*

Test Your Understanding: Chapter 9

Test yourself by filling in the blanks.

1. If f is a real-valued function of a real variable, then the domain and co-domain of f are both ____.

2. A point (x, y) lies on the graph of a real-valued function of a real variable f if, and only if, ____.

3. If a is any nonnegative real number, then the power function with exponent a, p_a, is defined by ____.

4. Given a function $f: \mathbf{R} \to \mathbf{R}$ and a real number M, the function Mf is defined by ____.

5. Given a function $f: \mathbf{R} \to \mathbf{R}$, to prove that f is increasing, you suppose that ____ and then you show that ____.

6. Given a function $f: \mathbf{R} \to \mathbf{R}$, to prove that f is decreasing, you suppose that ____ and then you show that ____.

7. A sentence of the form "$A|g(x)| \leq |f(x)|$ for all $x > a$," translates into Ω-notation as ____.

8. A sentence of the form "$|f(x)| \leq B|g(x)|$ for all $x > b$," translates into O-notation as ____.

9. A sentence of the form "$A|g(x)| \leq |f(x)| \leq B|g(x)|$ for all $x > k$," translates into Θ-notation as ____.

10. When $x > 1$, x^2 ____ x and x^5 ____ x^2.

11. According to the theorem on polynomial orders, if $p(x)$ is a polynomial in x, then $p(x)$ is $\Theta(x^n)$, where n is ____.

12. If n is a positive integer, then $1 + 2 + 3 + \cdots + n$ has order ____.

13. When an algorithm segment contains a nested **for-next** loop, you can find the number of times the loop will iterate by constructing a table in which each column represents ____.

14. In the worst case, the sequential search algorithm has to look through ____ elements of the input array before it terminates

15. The worst case order of the insertion sort algorithm is _____, and its average case order is _____.

16. The domain of the exponential function is _____, and its range is _____.

17. The domain of the logarithmic function is _____, and its range is _____.

18. If k is an integer and $2^k \leq x < 2^{k+1}$, then $\lfloor \log_2 x \rfloor =$ _____.

19. If b is a real number with $b > 1$ and if x is a sufficiently large real number, then when the quantities x, x^2, $\log_b x$, and $x \log_b x$ are arranged in order of increasing size, the result is _____.

20. If n is a positive integer, then $1 + \frac{1}{2} + \frac{1}{3} + \cdots + \frac{1}{n}$ has order _____.

21. To solve a problem using a divide-and-conquer algorithm, you reduce it to _____, which _____ and so forth until _____.

22. To search an array using the binary search algorithm in each step, you compare a middle element of the array to _____. If the middle element is less than _____, you _____, and if the middle element is greater than _____, you _____.

23. The worst case order of the binary search algorithm is _____.

24. To sort an array using the merge sort algorithm, in each step until the last one you split the array into approximately two equal sections and sort each section using _____. Then you _____ the two sorted sections.

25. The worst case order of the merge sort algorithm is _____.

Answers

1. sets of real numbers
2. $y = f(x)$
3. $p_a(x) = x^a$ for all real numbers x
4. $(Mf)(x) = M \cdot f(x)$ for $x \in \mathbf{R}$
5. x_1 and x_2 are any real numbers such that $x_1 < x_2$
   $f(x_1) < f(x_2)$
6. x_1 and x_2 are any real numbers such that $x_1 < x_2$
   $f(x_1) > f(x_2)$
7. $f(x)$ is $\Omega(g(x))$
8. $f(x)$ is $O(g(x))$
9. $f(x)$ is $\Theta(g(x))$
10. $>, >$
11. the degree of $p(x)$
12. n^2
13. one iteration of the innermost loop
14. n
15. n^2, n^2
16. the set of all real numbers, the set of all positive real numbers
17. the set of all positive real numbers, the set of all real numbers
18. k
19. $\log_b x \leq x \leq x \log_b x \leq x^2$
20. $\ln x$ (or, equivalently, $\log_2 x$)
21. a fixed number of smaller problems of the same kind
    can themselves be reduced to the same finite number of smaller problems of the same kind
    easily resolved problems are obtained

22. the element you are looking for
    the element you are looking for
    apply the binary search algorithm to the lower half of the array
    the element you are looking for
    apply the binary search algorithm to the upper half of the array

23. $\log_2 n$, where n is the length of the array
24. merge sort; merge
25. $n \log_2 n$

Chapter 10: Relations

The first section of this chapter is an introduction to concepts and notation with emphasis on learning equivalent ways to specify and represent relations, both finite and infinite. In Section 10.2 the reflexivity, symmetry, and transitivity properties of binary relations are introduced and explored, and in Section 10.3 equivalence relations are discussed. As you work on these sections, you will frequently use the fact that the same proof outline is used to prove and disprove universal conditional statements no matter what their mathematical context.

Section 10.4 deepens and extends the discussion of congruence relations in Sections 10.2 and 10.3 through applications to modular arithmetic and cryptography. The section is designed to make it possible to give you meaningful practice with RSA cryptography without having to spend several weeks on the topic. After a brief introduction to the idea of cryptography, the first part of the section is devoted to helping you develop the facility with modular arithmetic that is needed to perform the computations for RSA cryptography, especially finding least positive residues of integers raised to large positive powers and using the Euclidean algorithm to compute positive inverses modulo a number. Proofs of the underlying mathematical theory are left to the end of the section.

The last section of the chapter introduces another type of binary relation that is especially important in computer science: partial order relations

Section 10.1

3. *b.* *Proof*: Let n be any even integer. Then $n - 0 = n$ is also even, and so $n\,E\,0$ by definition of E.

6. *a.* Yes, $2 \geq 1$. Yes, $2 \geq 2$. No, $2 \not\geq 3$. Yes, $-1 \geq -2$.

9. *b.* No, $\{a\}$ has one element and $\{a, b\}$ has two.   *c.* Yes, both have one element.

12. *a.*

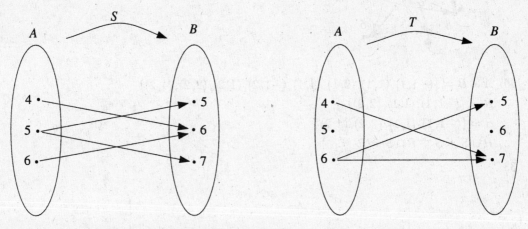

b. S is not a function because $(5, 5) \in S$ and $(5, 7) \in S$ and $5 \neq 7$. So S does not satisfy property (2) of the definition of function. T is not a function both because $(5, x) \notin T$ for any x in B and because $(6, 5) \in T$ and $(6, 7) \in T$ and $5 \neq 7$. So T does not satisfy either property (1) or property (2) of the definition of function.

15. *a.* There are 2^{mn} binary relations from A to B because a binary relation from A to B is any subset of $A \times B$, $A \times B$ is a set with mn elements (since A has m elements and B has n elements), and the number of subsets of a set with mn elements is 2^{mn} (by Theorem 5.3.1).

b. In order to define a function from A to B we must specify exactly one image in B for each of the m elements in A. So we can think of constructing a function from A to B as an m-step process, where step i is to choose an image for the ith element of A (for $i = 1, 2, \ldots, m$). Because there are n choices of image for each of the m elements, by the multiplication rule, the total number of functions is $\underbrace{n \cdot n \cdot n \cdots n}_{m\text{ factors}} = n^m$.

c. $\dfrac{n^m}{2^{nm}} = \left(\dfrac{n}{2^n}\right)^m$

18. $S = \{(3,6), (4,4), (5,5)\}$ $\quad$ $S^{-1} = \{(6,3), (4,4), (5,5)\}$

21. *b.* A function $F: X \to Y$ is onto if, and only if, for all $y \in Y$, $\exists x \in X$ such that $(x, y) \in F$.

24.

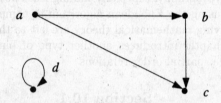

27.

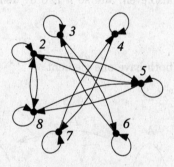

30. $A \times B = \{(-1,1), (1,1), (2,1), (4,1), (-1,2), (1,2), (2,2), (4,2)\}$
$R = \{(-1,1), (1,1), (2,2)\}$
$S = \{(-1,1), (1,1), (2,2), (4,2)\}$
$R \cup S = S$ $\quad$ $R \cap S = R$

33.

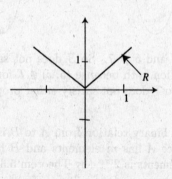 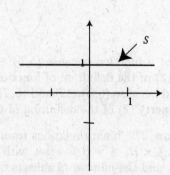

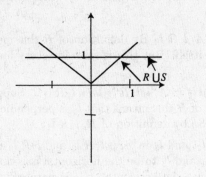

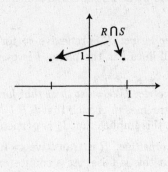

Section 10.2

24. *$\mathscr{P}$ is not reflexive:* $\mathscr{P}$ is reflexive $\Leftrightarrow$ for all sets $A \in \mathscr{P}(X)$, $A \; \mathscr{P} \; A$. By definition of $\mathscr{P}$ this means that for all sets A in $\mathscr{P}(X)$, $N(A) < N(A)$. But this is false for every set in $\mathscr{P}(X)$. For instance, let $A = \emptyset$. Then $N(A) = 0$, and 0 is not less than 0.

   $\mathscr{R}$ is not symmetric: For R to be symmetric would mean that for all sets A and B in $\mathscr{P}(X)$, if $A \; \mathscr{R} \; B$ then $B \; \mathscr{R} \; A$. By definition of $\mathscr{R}$, this would mean that for all sets A and B in $\mathscr{P}(X)$, if $N(A) < N(B)$, then $N(B) < N(A)$. But this is false for all sets A and B in $\mathscr{P}(X)$. For instance, take $A = \emptyset$ and $B = \{a\}$. Then $N(A) = 0$ and $N(B) = 1$. It follows that A is related to B by $\mathscr{R}$ (since $0 < 1$), but B is not related to A by $\mathscr{R}$ (since $1 \not< 0$).

   $\mathscr{R}$ is transitive: To prove transitivity of $\mathscr{R}$, we must show that for all sets A, B, and C in $\mathscr{P}(X)$, if $A \; \mathscr{R} \; B$ and $B \; \mathscr{R} \; C$ then $A \; \mathscr{R} \; C$. By definition of $\mathscr{R}$ this means that for all sets A, B, and C in $\mathscr{P}(X)$, if $N(A) < N(B)$ and $N(B) < N(C)$, then $N(A) < N(C)$. But this is true by the transitivity property of order (Appendix A, T17).

27. *$\mathscr{R}$ is not reflexive:* $\mathscr{R}$ is reflexive $\Leftrightarrow$ for all sets $X \in \mathscr{P}(A)$, $X \; \mathscr{R} \; X$. By definition of $\mathscr{R}$ this means that for all sets X in $\mathscr{P}(A)$, $X \neq X$. But this is false for every set in $\mathscr{P}(A)$. For instance, let $X = \emptyset$. It is not true that $\emptyset \neq \emptyset$.

   $\mathscr{R}$ is symmetric: $\mathscr{R}$ is symmetric $\Leftrightarrow$ for all sets X and Y in $\mathscr{P}(A)$, if $X \; \mathscr{R} \; Y$ then $Y \; \mathscr{R} \; X$. By definition of $\mathscr{R}$, this means that for all sets X and Y in $\mathscr{P}(A)$, if $X \neq Y$, then $Y \neq X$. But this is true.

   $\mathscr{R}$ is not transitive: $\mathscr{R}$ is transitive $\Leftrightarrow$ for all sets X, Y, and Z in $\mathscr{P}(A)$, if $X \; \mathscr{R} \; Y$ and $Y \; \mathscr{R} \; Z$ then $X \; \mathscr{R} \; Z$. By definition of $\mathscr{R}$ this means that for all sets X, Y, and Z in $\mathscr{P}(A)$, if $X \neq Y$ and $Y \neq Z$, then $X \neq Z$. But this is false as the following counterexample shows. Since $A \neq \emptyset$, there exists an element x in A. Let $X = \{x\}$, $Y = \emptyset$, and $Z = \{x\}$. Then $X \neq Y$ and $Y \neq Z$, but $X = Z$.

33. *R is reflexive:* R is reflexive $\Leftrightarrow$ for all points p in A, $p \; R \; p$. By definition of R this means that for all elements p in A, p and p both lie on the same half line emanating from the origin. But this is true.

   R is symmetric: *[We must show that for all points p_1 and p_2 in A, if $p_1 R \; p_2$ then $p_2 R \; p_1$.]* Suppose p_1 and p_2 are points in A such that $p_1 R \; p_2$. By definition of R this means that p_1 and p_2 lie on the same half line emanating from the origin. But this implies that p_2 and p_1 lie on the same half line emanating from the origin. So by definition of R, $p_2 R \; p_1$.

   R is transitive: *[We must show that for all points p_1, p_2 and p_3 in A, if $p_1 R \; p_2$ and $p_2 R \; p_3$ then $p_1 R \; p_3$.]* Suppose p_1, p_2, and p_3 are points in A such that $p_1 R \; p_2$ and $p_2 R \; p_3$. By definition of R, this means that p_1 and p_2 lie on the same half line emanating from the origin and p_2 and p_3 lie on the same half line emanating from the origin. Since two points determine

a line, it follows that both p_1 and p_3 lie on the same half line determined by the origin and p_2. Thus p_1 and p_3 lie on the same half line emanating from the origin, and so by definition of R, $p_1 R p_3$.

36. *R is not reflexive:* R is reflexive $\Leftrightarrow$ for all lines l in A, $l R l$. By definition of R this means that for all lines l in the plane, l is perpendicular to itself. But this is false for every line in the plane.

   R is symmetric: [We must show that for all lines l_1 and l_2 in A, if $l_1 R l_2$ then $l_2 R l_1$.] Suppose l_1 and l_2 are lines in A such that $l_1 R l_2$. By definition of R this means that l_1 is perpendicular to l_2. But this implies that l_2 is perpendicular to l_1. So by definition of R, $l_2 R l_1$.

   R is not transitive: R is transitive $\Leftrightarrow$ for all lines l_1, l_2, and l_3 in A, if $l_1 R l_2$ and $l_2 R l_3$ then $l_1 R l_3$. But this is false. As a counterexample, take l_1 and l_3 to be the horizontal axis and l_2 to be the vertical axis. Then $l_1 R l_2$ and $l_2 R l_3$ because the horizontal axis is perpendicular to the vertical axis and the vertical axis is perpendicular to the horizontal axis. But $l_1 \not R l_3$ because the horizontal axis is not perpendicular to itself.

39. Algorithm   Test for Symmetry

   [The input for this algorithm consists of a binary relation R defined on a set A which is represented as the one-dimensional array $a[1], a[2], \ldots, a[n]$. To test whether R is symmetric, the variable "answer" is initially set equal to "yes" and then each pair of elements $a[i]$ and $a[j]$ of A is examined in turn to see whether the condition "if $(a[i], a[j]) \in R$ then $(a[j], a[i]) \in R$" is satisfied. If not, then "answer" is set equal to "no", the while loop is not repeated, and processing terminates.]

   Input: n *[a positive integer]*, $a[1], a[2], \ldots, a[n]$ *[a one-dimensional array representing a set A]*, R *[a subset of $A \times A$]*

   Algorithm Body:

   $i := 1$, *answer* := "yes"

   while (*answer* = "yes" and $i \leq n$)

       $j := 1$

       while (*answer* = "yes" and $j \leq n$)

           if $(a[i], a[j]) \in R$ and $(a[j], a[i]) \notin R$ **then** *answer* := "no"

           $j := j + 1$

       end while

       $i := i + 1$

   end while

   Output: *answer [a string]*

42. The following is an easily understood algorithm to construct the transitive closure of a relation. In fact, by keeping the variable names in the statements in the inner loop the same and interchanging the variable names governing the two outer loops, the algorithm can be made significantly more efficient. The resulting algorithm, known as Warshall's algorithm, is discussed in most books on the design and analysis of algorithms. See, for example, *Computer Algorithms: Introduction to Design and Analysis* (3rd Edition) by Sara Baase and Allen Van Gelder, Reading, Massachusetts: Addison-Wesley Publishing Company, 1999.

Algorithm   Constructing a Transitive Closure

   [The input for this algorithm consists of a binary relation R defined on a set A which is represented as the one-dimensional array $a[1], a[2], \ldots, a[n]$. The transitive closure is constructed by modifying the procedure used to test for transitivity described in the answer to exercise 39.

Initially, the transitive closure, R^t, is set equal to R. Then a check is made through all triples of elements of A. In all cases for which $(a[i], a[j]) \in R^t$, $(a[j], a[k]) \in R^t$, and $(a[i], a[k]) \notin R^t$, the pair $(a[i], a[k])$ is added to R^t. After R^t has been enlarged in this way, it still may not equal the actual transitive closure of R because some of the added pairs may combine with pairs already present to necessitate the presence of additional pairs. So if any pairs have been added, an additional pass through all triples of elements of A is made. If after all triples of elements of A have been examined no pair has been added, the current relation R^t is transitive and equals the transitive closure of R.]

Input: n *[a positive integer]*, $a[1], a[2], \ldots, a[n]$ *[a one-dimensional array representing a set A]*, R *[a subset of $A \times A$]*

Algorithm Body:

$R^t := R$

$repeat := \text{``yes''}$

while $(repeat = \text{``yes''})$

   $repeat := \text{``no''}$

   $i := 1$

   while $(i \leq n)$

      $j := 1$

      while $(j \leq n)$

         $k := 1$

         while $(k \leq n)$

            if $(a[i], a[j]) \in R^t$ and $(a[j], a[k]) \in R^t$ and $(a[i], a[k]) \notin R^t$ **then do**

               $R^t := R^t \cup \{(a[i], a[k])\}$

               $repeat := \text{``yes''}$ **end do**

               $k := k + 1$

         end while

         $j := j + 1$

      end while

      $i := i + 1$

   end while

end while

Output: R^t *[a subset of $A \times A$]*

48. R_4 is irreflexive. R_4 is not asymmetric because $(1, 2) \in R_4$ and $(2, 1) \in R_4$. R_4 is intransitive.

51. R_7 is irreflexive. R_7 is asymmetric. R_7 is intransitive (by default).

Section 10.3

6. distinct equivalence classes: $\{0, 3, -3\}$, $\{1, 4, -2\}$, $\{2, 5, -1, -4\}$

9. distinct equivalence classes: $\{\emptyset, \{0\}, \{1, -1\}, \{-1, 0, 1\}\}$, $\{\{1\}, \{0, 1\}\}$, $\{\{-1\}, \{0, -1\}\}$

12. distinct equivalence classes: $\{0, 3, -3\}$, $\{1, 4, -2, 2, -5, 5, -1, -4\}$

15. *b. Proof*: Suppose that m and n are integers such that $m \equiv n \pmod{d}$. *[We must show that m mod $d = n$ mod d.]* By definition of congruence, $d \mid (m-n)$, and so by definition of divisibility $m - n = dk$ for some integer k. Let m mod $d = r$. Then $m = dl + r$ for some integer l. Since $m - n = dk$, then by substitution, $(dl + r) - n = dk$, or, equivalently, $n = d(l - k) + r$. Since $l - k$ is an integer, it follows by definition of *mod*, that n mod $d = r$ also, and so m mod $d = n$ mod d *[as was to be shown]*.

Suppose that m and n are integers such that m mod $d = n$ mod d. *[We must show that $m \equiv n$ $\pmod{d}$.]* Let $r = m$ mod $d = n$ mod d. Then by definition of *mod*, $m = dp + r$ and $n = dq + r$ for some integers p and q. By substitution, $m - n = (dp + r) - (dq + r) = d(p - q)$. Since $p - q$ is an integer, it follows that $d \mid (m - n)$, and so by definition of congruence, $m \equiv n \pmod{d}$.

18. (1) The solution given in Appendix B for exercise 15 in Section 10.2 showed that E is reflexive, symmetric, and transitive. Thus E is an equivalence relation.

21. (1) *Proof:*

S is reflexive because for each part x in P, x has the same part number and is shipped from the same supplier as x.

S is symmetric because for all parts x and y in P, if x has the same part number and is shipped from the same supplier as y then y has the same part number and is shipped from the same supplier as x.

S is transitive because for all parts x, y, and z in P, if x has the same part number and is shipped from the same supplier as y and y has the same part number and is shipped from the same supplier as z then x has the same part number and is shipped from the same supplier as z.

S is an equivalence relation because it is reflexive, symmetric, and transitive.

(2) There are as many distinct equivalence classes as there are distinct ordered pairs of the form (n, s) where n is a part number and s is a supplier name and s supplies a part with the number n. Each equivalence class consists of all parts that have the same part number and are shipped from the same supplier.

24. (1) *Proof:*

D is reflexive: Suppose m is any integer. Since $m^2 - m^2 = 0$ and $3 \mid 0$, we have that $3 \mid (m^2 - m^2)$. Consequently, m D m by definition of D.

D is symmetric: Suppose m and n are any integers such that m D n. By definition of D this means that $3 \mid (m^2 - n^2)$, and so, by definition of divisibility, $m^2 - n^2 = 3k$ for some integer k. Now $n^2 - m^2 = -(m^2 - n^2)$. Hence by substitution, $n^2 - m^2 = -(3k) = 3 \cdot (-k)$. It follows that $3 \mid (n^2 - m^2)$ by definition of divisibility (since $-k$ is an integer), and thus n D m by definition of D.

D is transitive: Suppose m, n and p are any integers such that m D n and n D p. By definition of D, this means that $3 \mid (m^2 - n^2)$ and $3 \mid (n^2 - p^2)$, and so, by definition of divisibility, $m^2 - n^2 = 3k$ for some integer k, and $n^2 - p^2 = 3l$ for some integer l. Now $m^2 - p^2 = (m^2 - n^2) + (n^2 - p^2)$. Hence by substitution, $m^2 - p^2 = 3k + 3l = 3(k + l)$. It follows that $3 \mid (m^2 - p^2)$ by definition of divisibility (since $k + l$ is an integer), and thus m D p by definition of D.

D is an equivalence relation because it is reflexive, symmetric, and transitive.

(2) Since $m^2 - n^2 = (m - n)(m + n)$ for all integers m and n, m D n $\Leftrightarrow$ $3 \mid (m - n)$ or $3 \mid (m + n)$. Then by examining cases, one sees that m D n $\Leftrightarrow$ for some integers k and l either $(m = 3k$ and $n = 3l)$ or $(m = 3k + 1$ and $n = 3l + 1)$ or $(m = 3k + 1$ and $n = 3l + 2)$ or $(m = 3k + 2$ and $n = 3l + 1)$ or $(m = 3k + 2$ and $n = 3l + 2)$. Therefore, there are two distinct equivalence classes $\{m \in \mathbf{Z} \mid m = 3k$ for some integer $k\}$ and $\{m \in \mathbf{Z} \mid m = 3k + 1$ or $m = 3k + 2$ for some integer $k\}$.

27. (1) *Proof:*

|| is reflexive because for each point l in A, l is parallel to l.

|| is symmetric because for all points l_1 and l_2 in A, if l_1 is parallel to l_2 then l_2 is parallel to l_1.

|| is transitive because for all points l_1, l_2, and l_3 in A, if l_1 is parallel to l_2 and l_2 is parallel to l_3 then l_1 is parallel to l_3.

|| is an equivalence relation because it is reflexive, symmetric, and transitive.

36. *Proof:* Suppose R is an equivalence relation on a set A, a and b are in A, and $a \in [b]$. By definition of class, $a \mathrel{R} b$. We must show that $[a] = [b]$. To show that $[a] \subseteq [b]$, suppose $x \in [a]$. *[We must show that $x \in [b]$.]* By definition of class, $x \mathrel{R} a$. By transitivity of R, since $x \mathrel{R} a$ and $a \mathrel{R} b$ then $x \mathrel{R} b$. Thus by definition of class, $x \in [b]$ *[as was to be shown]* . To show that $[b] \subseteq [a]$, suppose $x \in [b]$. *[We must show that $x \in [a]$.]* By definition of class, $x \mathrel{R} b$. But also $a \mathrel{R} b$, and so by symmetry, $b \mathrel{R} a$. Thus since R is transitive and since $x \mathrel{R} b$ and $b \mathrel{R} a$, then $x \mathrel{R} a$. Therefore, by definition of class, $x \in [a]$ *[as was to be shown]*. Since we have proved both subset relations $[a] \subseteq [b]$ and $[b] \subseteq [a]$, we conclude that $[a] = [b]$.

39. *b. Proof:* Suppose (a, b), (a', b'), (c, d), and (c', d') are any elements of A such that $[(a, b)] = [(a', b')]$ and $[(c, d)] = [(c', d')]$. By definition of the relation, a, a', c, and c' are integers and b, b', d, and d' are nonzero integers, and $ab' = a'b$ (*) and $cd' = c'd$ (**). We must show that $[(a, b)] \cdot [(c, d)] = [(a', b')] \cdot [(c', d')]$. By definition of the multiplication, this equation holds if, and only if, $[(ac, bd)] = [(a'c', b'd')]$. And by definition of the relation, this equation holds if, and only if, $ac \cdot b'd' = bd \cdot a'c'$. (***) But multiplying equations (*) and (**) gives $ab' \cdot cd' = a'b \cdot c'd$. And by the associative and commutative laws for real numbers, this equation is equivalent to (***). Hence $[(a, b)] \cdot [(c, d)] = [(a', b')] \cdot [(c', d')]$.

d. The identity element for multiplication is $[(1, 1)]$. To prove this, suppose (a, b) is any element of A. We must show that $[(a, b)] \cdot [(1, 1)] = [(a, b)]$. But by definition of the multiplication this equation holds if, and only if, $[(a \cdot 1, b \cdot 1)] = [(a, b)]$. By definition of the relation, this equation holds if, and only if, $a \cdot 1 \cdot b = b \cdot 1 \cdot a$, and this equation holds for all integers a and b. Thus $[(a, b)] \cdot [(1, 1)] = [(a, b)]$ *[as was to be shown]*.

f. Given any $(a, b) \in A$ with $a \neq 0$, $[(b, a)]$ is an inverse for multiplication for $[(a, b)]$. To prove this, we must show that $[(a, b)] \cdot [(b, a)] = [(1, 1)]$, which (by part (d)) is the identity element for multiplication. But by definition of the multiplication, $[(a, b)] \cdot [(b, a)] = [(ab, ba)]$. And by definition of the relation, $[(ab, ba)] = [(1, 1)]$ if, and only if, $ab \cdot 1 = ba \cdot 1$, which is true for all integers a and b. Thus $[(a, b)] \cdot [(b, a)] = [(1, 1)]$ *[as was to be shown]*.

42. *Proof:* Suppose R is a binary relation on a set A, R is symmetric and transitive, and for every x in A there is a y in A such that $x \mathrel{R} y$. Suppose x is any particular but arbitrarily chosen element of A. By hypothesis, there is a y in A such that $x \mathrel{R} y$. By symmetry, $y \mathrel{R} x$, and so by transitivity $x \mathrel{R} x$. Therefore, R is reflexive. Since we already know that R is symmetric and transitive, we conclude that R is an equivalence relation.

Section 10.4

6. *Proof:* Given any integer $n > 1$ and any integer a with $0 \leq a < n$, the notation $[a]$ denotes the equivalence class of a for the relation of congruence modulo n (Theorem 10.4.2). We first show that given any integer m, m is in one of the classes $[0], [1], [2], \ldots, [n-1]$. The reason is that, by the quotient-remainder theorem, $m = nk + a$, where k and a are integers and $0 \leq a < n$, and so, by Theorem 10.4.1, $m \equiv a \pmod{n}$. It follows by Lemma 10.3.2 that $[m] = [a]$. Next we use an argument by contradiction to show that all the equivalence classes $[0], [1], [2], \ldots, [n-1]$ are distinct. For suppose not. That is, suppose a and b are integers with

$0 \le a < n$ and $0 \le b < n$, $a \ne b$, and $[a] = [b]$. Without loss of generality, we may assume that $a > b \ge 0$, which implies that $-a < -b \le 0$. Adding a to all parts of the inequality gives $0 < a - b \le a$. By Theorem 10.3.4, $[a] = [b]$ implies that $a \equiv b \pmod{n}$. Hence, by Theorem 10.4.1, $n \mid (a - b)$, and so, by Example 3.3.3, $n \le a - b$.. But $a < n$. Thus $n \le a - b \le a < n$, which is contradictory. Therefore the supposition is false, and we conclude that all the equivalence classes $[0], [1], [2], \ldots, [n - 1]$ are distinct.

12. *b. Proof*: Suppose a is a positive integer. Then $a = \sum_{k=0}^{n} d_k 10^k$, for some nonnegative integer n and integers d_k where $0 \le d_k < 10$ for all $k = 1, 2, \ldots, n$. By Theorem 10.4.3,

$$a = \sum_{k=0}^{n} d_k 10^k \equiv \sum_{k=0}^{n} d_k \cdot 1 \equiv \sum_{k=0}^{n} d_k \pmod{9}$$

because, by part (a), each $10^k \equiv 1 \pmod{9}$. Hence, by Theorem 10.4.1, both a and $\sum_{k=0}^{n} d_k$ have the same remainder upon division by 9, and thus if either one is divisible by 9, so is the other.

18. $48^1 \bmod 713 = 48$

$48^2 \bmod 713 = 165$

$48^4 \bmod 713 = 165^2 \bmod 713 = 131$

$48^8 \bmod 713 = 131^2 \bmod 713 = 49$

$48^{16} \bmod 713 = 49^2 \bmod 713 = 262$

$48^{32} \bmod 713 = 262^2 \bmod 713 = 196$

$48^{64} \bmod 713 = 196^2 \bmod 713 = 627$

$48^{128} \bmod 713 = 627^2 \bmod 713 = 266$

$48^{256} \bmod 713 = 266^2 \bmod 713 = 169$

Hence, by Theorem 10.4.3, $48^{307} = 48^{256+32+16+2+1} = 48^{256}48^{32}48^{16}48^{2}48^{1} \equiv 169 \cdot 196 \cdot 262 \cdot 165 \cdot 48 \equiv 12 \pmod{713}$, and thus $48^{307} \bmod 713 = 12$.

21. The letters in EXCELLENT translate numerically into 05, 24, 03, 05,12, 12, 05, 14, 20. The solutions for exercises 19 and 20 in Appendix B and above show that E, L, and C are encrypted as 15, 23, and 27, respectively. To encrypt X, we compute $24^3 \bmod 55 = 19$, to encrypt N, we compute $14^3 \bmod 55 = 49$, and to encrypt T, we compute $20^3 \bmod 55 = 25$. So the ciphertext is 15 19 27 15 23 23 15 49 25.

24. By Example 10.4.10, the decryption key is 27. Thus the residues modulo 55 for 51^{27}, 14^{27}, 49^{27}, and 15^{27} must be found and then translated into letters of the alphabet. Because $27 = 16 + 8 + 2 + 1$, we first perform the following computations:

$51^1 \equiv 51 \pmod{55}$	$14^1 \equiv 14 \pmod{55}$	$49^1 \equiv 49 \pmod{55}$
$51^2 \equiv 16 \pmod{55}$	$14^2 \equiv 31 \pmod{55}$	$49^2 \equiv 36 \pmod{55}$
$51^4 \equiv 16^2 \equiv 36 \pmod{55}$	$14^4 \equiv 31^2 \equiv 26 \pmod{55}$	$49^4 \equiv 36^2 \equiv 31 \pmod{55}$
$51^8 \equiv 36^2 \equiv 31 \pmod{55}$	$14^8 \equiv 26^2 \equiv 16 \pmod{55}$	$49^8 \equiv 31^2 \equiv 26 \pmod{55}$
$51^{16} \equiv 31^2 \equiv 26 \pmod{55}$	$14^{16} \equiv 16^2 \equiv 36 \pmod{55}$	$49^{16} \equiv 26^2 \equiv 16 \pmod{55}$

Then

$51^{27} \bmod 55 = (26 \cdot 31 \cdot 16 \cdot 51) \bmod 55 = 6,$

$14^{27} \bmod 55 = (36 \cdot 16 \cdot 31 \cdot 14) \bmod 55 = 9,$

$49^{27} \bmod 55 = (16 \cdot 26 \cdot 36 \cdot 49) \bmod 55 = 14.$

In addition, we know from the solution to exercise 23 above that $15^{27} \bmod 55 = 5$. But 6, 9, 14, and 5 translate into letters as F, I, N, and E. So the message is FINE.

27. *Step 1:* $4158 = 1568 \cdot 2 + 1022$, and so $1022 = 4158 - 1568 \cdot 2$

    Step 2: $1568 = 1022 \cdot 1 + 546$, and so $546 = 1568 - 1022$

    Step 3: $1022 = 546 \cdot 1 + 476$, and so $476 = 1022 - 546$

    Step 4: $546 = 476 \cdot 1 + 70$, and so $70 = 546 - 476$

    Step 5: $476 = 70 \cdot 6 + 56$, and so $56 = 476 - 70 \cdot 6$

    Step 6: $70 = 56 \cdot 1 + 14$, and so $14 = 70 - 56$

    Step 7: $56 = 14 \cdot 4 + 0$, and so $\gcd(4158, 1568) = 14$,

    which is the remainder obtained just before the final division.

    Substitute back through steps 6–1:

    $$14 = 70 - 56 = 70 - (476 - 70 \cdot 6) = 70 \cdot 7 - 476$$
    $$= (546 - 476) \cdot 7 - 476 = 7 \cdot 546 - 8 \cdot 476$$
    $$= 7 \cdot 546 - 8 \cdot (1022 - 546) = 15 \cdot 546 - 8 \cdot 1022$$
    $$= 15 \cdot (1568 - 1022) - 8 \cdot 1022 = 15 \cdot 1568 - 23 \cdot 1022$$
    $$= 15 \cdot 1568 - 23 \cdot (4158 - 1568 \cdot 2) = 61 \cdot 1568 - 23 \cdot 4158$$

    (It is always a good idea to verify that no mistake has been made by verifying that the final expression really does equal the greatest common divisor. In this case, a computation shows that the answer is correct.)

30. *Proof:* Suppose a and b are positive integers, $S = \{x \mid x \text{ is a positive integer and } x = as + bt$ for some integers s and $t\}$, and c is the least element of S. We will show that $c \mid b$. By the quotient-remainder theorem, $b = cq + r$ (*) for some integers q and r with $0 \le r < c$. Now because c is in S, $c = as + bt$ for some integers s and t. Thus, by substitution into equation (*), $r = b - cq = b - (as + bt)q = a(-sq) + b(1 - tq)$. Hence, by definition of S, either $r = 0$ or $r \in S$. But if $r \in S$, then $r \ge c$ because c is the least element of S, and thus both $r < c$ and $r \ge c$ would be true, which would be a contradiction. Therefore, $r \notin S$, and thus by elimination, we conclude that $r = 0$. It follows that $b - cq = 0$, or, equivalently, $b = cq$, and so $c \mid b$ *[as was to be shown]*.

33. *Proof:* Suppose a, b, and c are integers such that $\gcd(a, b) = 1$, $a \mid c$, and $b \mid c$. We will show that $ab \mid c$. By Corollary 10.4.6 (or by Theorem 10.4.5), $as + bt = 1$. Also, by definition of divisibility, $c = au = bv$, for some integers u and v. Hence, by substitution, $c = asc + btc = as(bv) + bt(au) = ab(sv + tu)$. But $sv + tu$ is an integer, and so, by definition of divisibility, $ab \mid c$ *[as was to be shown]*.

42. *b.* When $a = 8$ and $p = 11$, $a^{p-1} = 8^{10} = 1 \equiv 10731741824 \pmod{11}$ because $10731741824 - 1 = 11 \cdot 97612893$.

45. To solve this problem, we need to find a positive integer x such that $x \equiv 2 \pmod{15}$, $x \equiv 1 \pmod{14}$, and $x \equiv 0 \pmod{13}$. We apply the technique in the proof of the Chinese remainder theorem with $n_1 = 15$, $n_2 = 14$, $n_3 = 13$, $a_1 = 2$, $a_2 = 1$, and $a_3 = 0$. Then $N = 15 \cdot 14 \cdot 13 = 2730$, $N_1 = 14 \cdot 13 = 182$, $N_2 = 15 \cdot 13 = 195$, and $N_3 = 15 \cdot 14 = 210$.

    To find x_1, we solve $N_1 x_1 = 182 x_1 \equiv 1 \pmod{15}$. Now $182 = 15 \cdot 12 + 2$, and so $2 = 182 - 15 \cdot 12$. Also $15 = 2 \cdot 7 + 1$, and so $1 = 15 - 2 \cdot 7$. Hence, by substitution, $1 = 15 - (182 - 15 \cdot 12) \cdot 7 = 182 \cdot (-7) + 85 \cdot 15$, and so $182 \cdot (-7) \equiv 1 \pmod{15}$. Thus $x_1 \equiv -7 \pmod{15} \equiv 8 \pmod{15}$ because $15 \mid ((-7) - 8)$. So we may take $x_1 = 8$.

    To find x_2, we solve $N_2 x_2 = 195 x_2 \equiv 1 \pmod{14}$. Now $195 = 14 \cdot 13 + 13$, and so $13 = 195 - 14 \cdot 13$. Also $14 = 1 \cdot 13 + 1$, and so $1 = 14 - 13$. Hence, by substitution, $1 = 14 - (195 - 14 \cdot 13) = 195 \cdot (-1) + 14 \cdot 14$, and so $195 \cdot (-1) \equiv 1 \pmod{14}$. Thus $x_2 \equiv -1 \pmod{14} \equiv 13 \pmod{14}$ because $14 \mid (-1 - 13)$. So we may take $x_2 = 13$.

To find x_3, we solve $N_3 x_3 = 210 x_3 \equiv 0 \pmod{13}$. Now $210 = 13 \cdot 16 + 2$, and so $2 = 210 - 13 \cdot 16$. Also $13 = 2 \cdot 6 + 1$, and so $1 = 13 - 2 \cdot 6$. Hence, by substitution, $1 = 13 - (210 - 13 \cdot 16) \cdot 6 = 210 \cdot (-6) + 97 \cdot 13$, and so $210 \cdot (-6) \equiv 1 \pmod{13}$. Thus $x_3 \equiv -6 \pmod{13} \equiv 7 \pmod{13}$ because $13 \mid (-6 - 7)$. So we may take $x_2 = 7$. (Strictly speaking, as you will see below, we did not need to calculate x_3 because $a_3 = 0$.)

By the proof of the Chinese remainder theorem, a solution x for the congruences is $x = a_1 N_1 x_1 + a_2 N_2 x_2 + a_3 N_3 x_3 = 2 \cdot 182 \cdot 8 + 1 \cdot 195 \cdot 13 + 0 \cdot 210 \cdot 7 = 5447$. But $5447 \bmod 2730 = 2717$. Thus the least positive solution to the system of congruences is 2717. To check this answer, observe that $2717 = 15 \cdot 181 + 2$, $2717 = 14 \cdot 194 + 1$, and $2717 = 13 \cdot 209$.

Section 10.5

3. R is not antisymmetric. *Counterexample*: Let $s = 0$ and $t = 1$. Then $s \, R \, t$ and $t \, R \, s$ because $l(s) \leq l(t)$ and $l(t) \leq l(s)$, since both $l(s)$ and $l(t)$ equal 1, but $s \neq t$.

6. R is a partial order relation.

   Proof:

   R is reflexive: Suppose $r \in P$. Then $r = r$, and so by definition of R, $r \, R \, r$.

   R is antisymmetric: Suppose $r, s \in P$ and $r \, R \, s$ and $s \, R \, r$. *[We must show that $r = s$.]* By definition of R, either r is an ancestor of s or $r = s$ and either s is an ancestor of r or $s = r$. Now it is impossible for both r to be an ancestor of s and s to be an ancestor of r. Hence one of these conditions must be false, and so $r = s$ *[as was to be shown]*.

   R is transitive: Suppose $r, s, t \in P$ and $r \, R \, s$ and $s \, R \, t$. *[We must show that $r \, R \, t$.]* By definition of R, either r is an ancestor of s or $r = s$ and either s is an ancestor of t or $s = t$. In case r is an ancestor of s and s is an ancestor of t, then r is an ancestor of t, and so $r \, R \, t$. In case r is an ancestor of s and $s = t$, then r is an ancestor of t, and so $r \, R \, t$. In case $r = s$ and s is an ancestor of t, then r is an ancestor of t, and so $r \, R \, t$. In case $r = s$ and $s = t$, then $r = t$, and so $r \, R \, t$. Thus in all four possible cases, $r \, R \, t$ *[as was to be shown]*.

   Since R is reflexive, antisymmetric, and transitive, R is a partial order relation.

9. R is not a partial order relation because R is not antisymmetric. *Counterexample*: Let $x = 2$ and $y = -2$. Then $x \, R \, y$ because $(-2)^2 \leq 2^2$, and $y \, R \, x$ because $2^2 \leq (-2)^2$. But $x \neq y$ because $2 \neq -2$.

12. *Proof*:

    $\preceq$ is reflexive: Suppose s is in S. If $s = \epsilon$, then $s \preceq s$ by (3). If $s \neq \epsilon$, then $s \preceq s$ by (1). Hence in either case, $s \preceq s$.

    $\preceq$ is antisymmetric: Suppose s and t are in S and $s \preceq t$ and $t \preceq s$. *[We must show that $s = t$.]* By definition of S, either $s = \epsilon$ or $s = a_1 a_2 \ldots a_m$ and either $t = \epsilon$ or $t = b_1 b_2 \ldots b_n$ for some positive integers m and n and elements $a_1, a_2, \ldots, a_m$ and $b_1, b_2, \ldots, b_n$ in A. It is impossible to have $s \preceq t$ by virtue of condition (2) because in that case there is no condition that would give $t \preceq s$. *[For suppose $s \preceq t$ by virtue of condition (2). Then for some integer k with $k \leq m$, $k \leq n$, and $k \geq 1$, $a_i = b_i$ for all $i = 1, 2, \ldots, k - 1$, and $a_k \, R \, b_k$ and $a_k \neq b_k$. In this situation, it is clearly impossible for $t \preceq s$ by virtue either of condition (1) or (3), and so if $t \preceq s$, then it must be by virtue of condition (2). But in that case, since $a_k \neq b_k$, it must follow that $b_k \, R \, a_k$, and so since R is a partial order relation, $a_k = b_k$. However, this contradicts the fact that $a_k \neq b_k$. Hence it cannot be the case that $s \preceq t$ by virtue of condition (2).]* Similarly, it is impossible for $t \preceq s$ by virtue of condition (2). Hence $s \preceq t$ and $t \preceq s$ by virtue either of condition (1) or of condition (3). In case $s \preceq t$ by virtue of condition (1), then neither s nor t is the null string and so $t \preceq s$ by virtue of condition (1) also. Then by (1)

$m \leq n$ and $a_i = b_i$ for all $i = 1, 2, \ldots, m$ and $n \leq m$ and $b_i = a_i$ for all $i = 1, 2, \ldots, m$, and so in this case $s = t$. In case $s \preceq t$ by virtue of condition (3), then $s = \epsilon$, and so since $t \preceq s$, $t \preceq \epsilon$. But the only condition that can give this result is (3) with $t = \epsilon$. Hence in this case, $s = t = \epsilon$. Thus in all possible cases, if $s \preceq t$ and $t \preceq s$, then $s = t$ [as was to be shown].

$\preceq$ is transitive: Suppose s and t are in S and $s \preceq t$ and $t \preceq u$. [We must show that $s \preceq u$.] By definition of S, either $s = \epsilon$ or $s = a_1 a_2 \ldots a_m$, either $t = \epsilon$ or $t = b_1 b_2 \ldots b_n$, and either $u = \epsilon$ or $u = c_1 c_2 \ldots c_p$ for some positive integers m, n, and p and elements $a_1, a_2, \ldots, a_m$, $b_1, b_2, \ldots, b_n$, and $c_1, c_2, \ldots, c_p$ in A.

Case 1 $(s = \epsilon)$: In this case, $s \, R \, u$ by (3).

Case 2 $(s \neq \epsilon)$: In this case, since $s \, R \, t$, $t \neq \epsilon$ either, and since $t \, R \, u$, $u \neq \epsilon$ either.

Subcase a $(s \, R \, t$ and $t \, R \, u$ by condition (1)): Then $m \leq n$ and $n \leq p$ and $a_i = b_i$ for all $i = 1, 2, \ldots, m$ and $b_j = c_j$ for all $j = 1, 2, \ldots, n$. It follows that $a_i = c_i$ for all $i = i, 2, \ldots, m$, and so by (1), $s \, R \, u$.

Subcase b $(s \, R \, t$ by condition (1) and $t \, R \, u$ by condition (2)): Then $m \leq n$ and $a_i = b_i$ for all $i = 1, 2, \ldots, m$, and for some integer k with $k \leq n$, $k \leq p$, and $k \geq 1$, $b_j = c_j$ for all $j = 1, 2, \ldots, k - 1$, $b_k \, R \, c_k$, and $b_k \neq c_k$. If $k \leq m$, then s and u satisfy condition (2) [because $a_i = b_i$ for all $i = 1, 2, \ldots, m$ and so $k \leq m$, $k \leq p$, $k \geq 1$, $a_i = b_i = c_i$ for all $i = 1, 2, \ldots, k - 1$, $a_k \, R \, c_k$, and $a_k \neq c_k$]. If $k > m$, then s and u satisfy condition (1) [because $a_i = b_i = c_i$ for all $i = 1, 2, \ldots, m$]. Thus in either case $s \, R \, u$.

Subcase c $(s \, R \, t$ by condition (2) and $t \, R \, u$ by condition (1)): Then for some integer k with $k \leq m$, $k \leq n$, $k \geq 1$, $a_i = b_i$ for all $i = 1, 2, \ldots, k - 1$, $a_k \, R \, b_k$, and $a_k \neq b_k$, and $n \leq p$ and $b_j = c_j$ for all $j = i, 2, \ldots, n$. Then s and u satisfy condition (2) [because $k \leq n$, $k \leq p$ (since $k \leq n$ and $n \leq p$), $k \geq 1$, $a_i = b_i = c_i$ for all $i = 1, 2, \ldots, k - 1$ (since $k - 1 < n$), $a_k \, R c_k$ (since $b_k = c_k$ because $k \leq n$), and $a_k \neq c_k$ (since $b_k = c_k$ and $a_k \neq b_k$)]. Thus $s \, R \, u$.

Subcase d $(s \, R \, t$ by condition (2) and $t \, R \, u$ by condition (2)): Then for some integer k with $k \leq m$, $k \leq n$, $k \geq 1$, $a_i = b_i$ for all $i = 1, 2, \ldots, k - 1$, $a_k \, R \, b_k$, and $a_k \neq b_k$, and for some integer l with $l \leq n$, $l \leq p$, and $l \geq 1$, $b_j = c_j$ for all $j = 1, 2, \ldots, l - 1$, $b_l \, R \, c_l$, and $b_l \neq c_l$. If $k < l$, then $a_i = b_i = c_i$ for all $i = 1, 2, \ldots, k - 1$, $a_k \, R \, b_k$, $b_k = c_k$ (in which case $a_k \, R \, c_k$), and $a_k \neq c_k$ (since $a_k \neq b_k$). Thus if, $k < l$, then $s \preceq u$ by condition (2). If $k = l$, then $b_k \, R \, c_k$ (in which case $a_k \, R \, c_k$ by transitivity of R) and $b_k \neq c_k$. It follows that $a_k \neq c_k$ [for if $a_k = c_k$, then $a_k \, R \, b_k$ and $b_k \, R \, a_k$, which implies that $a_k = b_k$ (since R is a partial order) and contradicts the fact that $a_k \neq b_k$]. Thus if $k = l$, then $s \preceq u$ by condition (2). If $k > l$, then $a_i = b_i = c_i$ for all $i = 1, 2, \ldots, l - 1$, $a_l \, R \, c_l$ (because $b_l \, R \, c_l$ and $a_l = b_l$), $a_l \neq c_l$ (because $b_l \neq c_l$ and $a_l = b_l$). Thus if $k > l$, then $s \preceq u$ by condition (2). Hence in all cases $s \preceq u$.

The above arguments show that in all possible cases, $s \preceq u$ [as was to be shown]. Hence $\preceq$ is transitive.

Since $\preceq$ is reflexive, antisymmetric, and transitive, $\preceq$ is a partial order relation.

15. *Proof*: Suppose R is a relation on a set A and R is reflexive, symmetric, transitive, and anti-symmetric. We will show that R is the identity relation on A. First note that for all x and y in A, if $x \, R \, y$ then, because R is symmetric, $y \, R \, x$. But then, because R is also anti-symmetric $x = y$. Thus for all x and y in A, if $x \, R \, y$ then $x = y$. This argument, however, does not prove that R is the identity relation on A because the conclusion would also follow from the hypothesis (by default) in the case where $A \neq \emptyset$ and $R = \emptyset$. But when $A \neq \emptyset$, it is impossible for R to equal $\emptyset$ because R is reflexive, which means that $x \, R \, x$ for every x in A. Thus every element in A is related by R to itself, and no element in A is related to anything other than itself. It follows that R is the identity relation on A.

27. greatest element: (1,1)   least element: (0,0)

maximal elements: (1,1)   minimal elements: (0,0)

30. *c.* no greatest element and no least element

    d. greatest element: 9   least element: 1

33. *A* is not totally ordered by the given relation because $9 \nmid 12$ and $12 \nmid 9$.

42. *a. Proof*: Suppose *A* is any finite partially ordered set, ordered with respect to a relation $\preceq$, and *a* and *b* are greatest elements of *A*. By definition of greatest element, $x \preceq a$ for all *x* in *A*; in particular, $b \preceq a$. Similarly, $x \preceq b$ for all *x* in *A*, and so $a \preceq b$. Since $\preceq$ *is* a partial order, it is antisymmetric, and thus $a = b$. Hence *A* has at most one greatest element.

    b. Proof: Suppose *A* is any finite partially ordered set, ordered with respect to a relation $\preceq$, and *a* and *b* are least elements of *A*. By definition of least element, $a \preceq x$ for all *x* in *A*; in particular, $a \preceq b$. Similarly, $b \preceq x$ for all *x* in *A*, and so $b \preceq a$. Since $\preceq$ *is* a partial order, it is antisymmetric, and thus $a = b$. Hence *A* has at most one least element.

48. One such total order is $(0,0,0),(0,0,1),(0,1,0),(0,1,1),(1,0,0),(1,0,1),(1,1,0),(1,1,1)$.

51. *b.* (i) Annotate the given Hasse diagram by indicating in boxes the least number of days needed to accomplish each job, taking into account the time needed to perform prerequisite jobs.

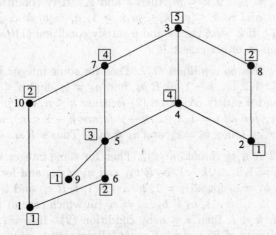

Therefore, at least five days are needed to perform all ten jobs.

(ii) At most four jobs can be performed at the same time. For instance, 10, 9, 6 and 2 could be performed simultaneously. One way to see why the maximum cannot be greater than four is to observe that *S* can be written as a union of four chains: $1 \preceq 10 \preceq 7 \preceq 3$, $1 \preceq 6 \preceq 5 \preceq 7 \preceq 3$, $9 \preceq 5 \preceq 4 \preceq 3$, and $2 \preceq 8 \preceq 3$, and at most one job from each chain can be performed at any one time.

General Review for Chapter 10

Definitions: How are the following terms defined?

- binary relation from a set A to a set B *(p. 572)*
- inverse of a binary relation from a set A to a set B *(p. 578)*
- n-ary relation R on $A_1 \times A_2 \times \cdots \times A_n$ *(p. 581)*
- reflexive, symmetric, and transitive properties of a binary relation *(p. 584)*
- transitive closure of a relation *(p. 588)*
- equivalence relation on a set *(p. 597)*
- equivalence class *(p. 599)*
- a is congruent to b modulo d *(p. 597)*
- plaintext and cyphertext *(p. 611)*
- residue of a modulo n *(p. 614)*
- d is a linear combination of a and b *(p. 619)*
- a and b are relatively prime *(p. 621)*
- an inverse of a modulo n *(p. 622)*
- antisymmetric binary relation *(p. 632)*
- partial order relation *(p. 634)*
- a and b are comparable *(p. 639)*
- total order relation *(p. 639)*
- chain, length of a chain *(p. 640)*
- maximal element, greatest element, minimal element, least element *(p. 641)*
- topological sorting *(p. 642)*

General Binary Relations

- Given the definition of a binary relation as a subset of a Cartesian product, what does it mean for one element to be related to another? *(p. 572)*
- How do you draw an arrow diagram for a binary relation? *(p. 574)*
- A function f from A to B is a binary relation from A to B that satisfies what special properties? *(p. 575)*
- Given a binary relation on a set, how do you draw a directed graph for the relation? *(p. 580)*

Properties of Binary Relations and Equivalence Relations

- How do you show that a binary relation on a finite set is reflexive? symmetric? transitive? *(p. 585)*
- How do you show that a binary relation on an infinite set is reflexive? symmetric? transitive? *(p. 589-92)*
- How do you show that a binary relation on a set is not reflexive? not symmetric? not transitive? *(p. 585, p. 590)*
- How do you find the transitive closure of a relation? *(p. 588)*
- What is the binary relation induced by a partition of a set? *(p. 595)*
- How do you prove basic properties of equivalence classes? *(p. 602)*
- Given an equivalence relation on a set A, what is the relationship between the distinct equivalence classes of the relation and the set A? *(p. 603)*
- In what way are rational numbers equivalence classes? *(p. 607)*

Cryptography

- How does the Caesar cipher work? *(p. 611)*
- If a, b, and n are integers with $n > 1$, what are some different ways of expressing the fact that $n \mid (a - b)$? *(p. 613)*
- If n is an integer with $n > 1$, is congruence modulo n an equivalence relation on the set of all integers? *(p. 614)*
- How do you add, subtract, and multiply integers modulo an integer $n > 1$? *(p. 615)*
- What is an efficient way to compute a^k where a is an integer with $a > 1$ and k is a large integer? *(p. 618)*
- How do you express the greatest common divisor of two integers as a linear combination of the integers? *(p. 620)*
- When can you find an inverse modulo n for a positive integer a, and how do you find it? *(p. 621)*
- How do you encrypt and decrypt messages using RSA cryptography? *(p. 624)*
- What is Euclid's lemma? How is it proved? *(p. 625)*
- What is Fermat's little theorem? How is it proved? *(p. 626)*
- What is the Chinese remainder theorem? How is it proved? *(p. 627)*
- Why does the RSA cipher work? *(p. 628)*

Partial Order Relations

- How do you show that a relation on a set is or is not antisymmetric? *(pp. 632-4)*
- If A is a set with a partial order relation R, S is a set of strings over A, and a and b are in S, how do you show that $a \preceq b$, where $\preceq$ denotes the lexicographic ordering of S? *(p. 636)*
- How do you construct the Hasse diagram for a partial order relation? *(p. 637)*
- How do you find a chain in a partially ordered set? *(p. 640)*
- Given a set with a partial order, how do you construct a topological sorting for the elements of the set? *(p. 642)*
- Given a job scheduling problem consisting of a number of tasks, some of which must be completed before others can be begun, how can you use a partial order relation to determine the minimum time needed to complete the job? *(p. 644)*

Test Your Understanding: Chapter 10

Test yourself by filling in the blanks.

1. A binary relation R from A to B is _____.

2. If R is a binary relation, the notation xRy means that _____.

3. If R is a binary relation, the notation $x \not{R} y$ means that _____.

4. For a binary relation R on a set A to be reflexive means that _____.

5. For a binary relation R on a set A to be symmetric means that _____.

6. For a binary relation R on a set A to be transitive means that _____.

7. To show that a binary relation R on an infinite set A is reflexive, you suppose that _____ and you show that _____.

8. To show that a binary relation R on an infinite set A is symmetric, you suppose that _____ and you show that _____.

9. To show that a binary relation R on an infinite set A is transitive, you suppose that _____ and you show that _____.

10. To show that a binary relation R on a set A is not reflexive, you _____.

11. To show that a binary relation R on a set A is not symmetric, you _____.

12. To show that a binary relation R on a set A is not transitive, you _____

13. Given a binary relation R on a set A, the transitive closure of R is the binary relation R^t on A that satisfies the following three properties: _____, _____, and _____.

14. For a binary relation on a set to be an equivalence relation, it must be _____.

15. The notation $m \equiv n \pmod{d}$ is read _____ and means that _____.

16. Given an equivalence relation R on a set A and given an element a in A, the equivalence class of a is denoted _____ and is defined to be _____.

17. If A is a set, R is an equivalence relation on A, and a and b are elements of A, then either $[a] = [b]$ or _____.

18. If A is a set and R is an equivalence relation on A, then the distinct equivalence classes of R form _____.

19. Let $A = \mathbf{Z} \times (\mathbf{Z} - \{0\})$, and define a binary relation R on A by specifying that for all (a, b) and (c, d) in A, $(a, b)R(c, d)$ if, and only if, $ad = bc$. Then there is exactly one equivalence class of R for each _____.

20. When letters of the alphabet are encrypted using the Caesar cipher, the encrypted version of a letter is _____.

21. If a, b, and n are integers with $n > 1$, the following are all different ways of expressing the fact that $n \mid (a - b)$: _____, _____, _____, _____.

22. If a, b, c, d, m and n are integers with $n > 1$ and if $a \equiv c \pmod{n}$ and $b \equiv d \pmod{n}$, then $a + b \equiv$ _____, $a - b \equiv$ _____, $ab \equiv$ _____, and $a^m \equiv$ _____.

23. If a, n, and k are positive integers with $n > 1$, an efficient way to compute $a^k \pmod{n}$ is to write k as a _____ and use the facts about computing products and powers modulo n.

24. To express a greatest common divisor of two integers as a linear combination of the integers, you use the extended _____ algorithm.

25. To find an inverse for a positive integer a modulo an integer n with $n > 1$, you express the number 1 as _____.

26. To encrypt a message M using RSA cryptography with public key pq and e, you use the formula _____, and to decrypt a message C, you use the formula _____, where _____.

27. Euclid's lemma says that for all integers a, b, and c if $\gcd(a, c) = 1$ and $a \mid bc$, then _____.

28. Fermat's little theorem says that if p is any prime number and a is any integer such that $p \nmid a$ then _____.

29. The Chinese remainder theorem says that if $n_1, n_2, \ldots, n_k$ are pairwise relatively prime positive integers and $a_1, a_2, \ldots, a_k$ are any integers, then the congruences $x \equiv a_i \pmod{n_i}$ for $i = 1, 2, \ldots, k$, have a _____.

30. The crux of the proof that the RSA cipher works is that if (1) p and q are large prime numbers, (2) $M < pq$, (3) M is relatively prime to pq, (4) e is relatively prime to $(p-1)(q-1)$, and (5) d is a positive inverse for e modulo $(p-1)(q-1)$, then $M \equiv$ ____.

31. For a binary relation R on a set A to be antisymmetric means that ____.

32. To show that a binary relation R on an infinite set A is antisymmetric, you suppose that ____ and you show that ____.

33. To show that a binary relation R on a set A is not antisymmetric, you ____.

34. To construct a Hasse diagram for a partial order relation, you start with a directed graph of the relation in which all arrows point upward and you eliminate ____, ____, and ____.

35. If A is a set that is partially ordered with respect to a relation $\preceq$ and if a and b are elements of A, we say that a and b are comparable if, and only if, ____ or ____.

36. A relation $\preceq$ on a set A is a total order if, and only if, ____.

37. If A is a set that is partially ordered with respect to a relation $\preceq$, and if B is a subset of A, then B is a chain if, and only if, for all a and b in B, ____.

38. Let A be a set that is partially ordered with respect to a relation $\preceq$, and let a be an element of A.

    (a) a is maximal if, and only if, ____.

    (b) a is a greatest element of A if, and only if, ____.

    (c) a is called minimal if, and only if, ____.

    (d) a is called a least element of A if, and only if, ____.

39. Given a set A that is partially ordered with respect to a relation $\preceq$, the relation $\preceq'$ is a topological sorting for $\preceq$, if, and only if, $\preceq'$ is a ____ and for all a and b in A if $a \preceq b$ then ____.

40. PERT and CPM are used to produce efficient ____.

Answers

1. a subset of $A \times B$
2. x is related to y by R
3. x is not related to y by R
4. for all x in A; $x\,R\,x$
5. for all x and y in A, if $x\,R\,y$ then $y\,R\,x$
6. for all x, y, and z in A, if $x\,R\,y$ and $y\,R\,z$ then $x\,R\,z$
7. x is any element of A; $x\,R\,x$
8. x and y are any elements of A such that $x\,R\,y$; $y\,R\,x$
9. x, y, and z are any elements of A such that $x\,R\,y$ and $y\,R\,z$; $x\,R\,z$
10. show the existence of an element x in A such that $x\,\not\!R\,x$
11. show the existence of elements x and y in A such that $x\,R\,y$ but $y\,\not\!R\,x$
12. show the existence of elements x, y, and z in A such that $x\,R\,y$ and $y\,R\,z$ but $x\,\not\!R\,z$
13. R^t is transitive; $R \subseteq R^t$; If S is any other transitive relation that contains R, then $R^t \subseteq S$
14. reflexive, symmetric, and transitive
15. m is congruent to n modulo d; d divides $m - n$
16. $[a]$; the set of all x in A such that $x\,R\,a$

17. $[a] \cap [b] = \emptyset$
18. a partition of A
19. rational number
20. three places in the alphabet to the right of the letter, with X wrapped around to A, Y to B, and Z to C
21. $a \equiv b \pmod{n}$
    $a = b + kn$ for some integer k
    a and b have the same nonnegative remainder when divided by n
    $a \bmod n = b \bmod n$
22. $(c + d) \pmod{n}$; $(c - d) \pmod{n}$; $(cd) \pmod{n}$; $c^m \pmod{n}$
23. sum of powers of 2
24. version of the Euclidean
25. a linear combination of a and n
26. $C = M^e \bmod pq$; $M = C^d \bmod pq$; d is a positive inverse for e modulo $(p-1)(q-1)$
27. $a \mid b$
28. $a^{p-1} \equiv 1 \pmod{p}$
29. simultaneous solution x that is unique modulo n, where $n = n_1 n_2 \cdots n_k$
30. $M^{ed} \pmod{pq}$
31. for all a and b in A, if $a \, R \, b$ and $b \, R \, a$ then $a = b$
32. a and b are any elements of A with $a \, R \, b$ and $b \, R \, a$; $a = b$
33. show the existence of elements a and b in A such that $a \, R \, b$ and $b \, R \, a$ and $a \neq b$
34. all loops; all arrows whose existence is implied by the transitive property; the direction indicators on the arrows
35. $a \preceq b$; $b \preceq a$
36. for any two elements a and b in A; either $a \preceq b$ or $b \preceq a$
37. a and b are comparable
38.
    (a) for all b in A either $b \preceq a$ or b and a are not comparable
    (b) for all b in A, $b \preceq a$
    (c) for all b in A either $a \preceq b$ or b and a are not comparable
    (d) for all b in A, $a \preceq b$

39. total order; $a \preceq b$
40. scheduling of tasks

Chapter 11: Graphs and Trees

The first section of this chapter introduces the terminology of graph theory, illustrating it in a variety of different instances. Several exercises are designed to clarify the distinction between a graph and a drawing of a graph. Other exercises explore the use of graphs to solve problems of various sorts. In some cases, you may be able to solve the given problems, such as the wolf, the goat, the cabbage and the ferryman, more easily without using graphs than using them. The point is that such problems *can* be solved using graphs and that for more complex problems involving, say, hundreds of possible states, a graphical representation coupled with a computer path-finding algorithm makes it possible find a solution that could not be discovered by trial-and-error alone. The variety of solutions for exercise 33, on the number of edges of a complete graph illustrates the relations among different branches of discrete mathematics. The rest of the exercises in this section are intended to give you practice in applying the theorem that relates the total degree of a graph to the number of its edges, especially for exploring properties of simple graphs, complete graphs, and bipartite graphs.

In Section 11.2 the general topic of paths and circuits is discussed, including the notion of connectedness and Euler and Hamiltonian circuits. As throughout the chapter, an attempt is made to balance the presentation of theory and application.

Section 11.3 introduces the concept of the adjacency matrix of a graph. The main theorem of the section states that the ijth entry of the kth power of the adjacency matrix equals the number of walks of length k from the ith to the jth vertices in the graph. Matrix multiplication is defined and explored in this section in a way that is intended to be adequate even if you have never seen the definition before.

The concept of graph isomorphism is discussed in Section 11.4. In this section the main theorem gives a list of isomorphic invariants that can be used to determine the non-isomorphism of two graphs.

The last two sections of the chapter deal with the subject of trees. Section 11.5 contains definitions, examples, and theorems giving necessary and sufficient conditions for graphs to be trees, and it also includes the definition of rooted tree, binary tree, and the theorems that relate the number of internal to the number of terminal vertices of a full binary tree and the maximum height of a binary tree to the number of its terminal vertices. Section 11.6 on spanning trees contains Kruskal's and Prim's algorithms and proofs of their correctness, as well as applications of minimum spanning trees.

Section 11.1

6.

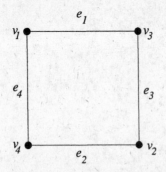

9. (i) e_1, e_2, e_7 are incident on e_1.

   (ii) v_1 and v_2 are adjacent to v_3.

(iii) e_2 and e_7 are adjacent to e_1.

(iv) e_1 and e_3 are loops.

(v) e_4 and e_5 are parallel.

(vi) v_4 is an isolated vertex.

(vii) degree of $v_3 = 2$

(viii) total degree of the graph $= 14$

18.

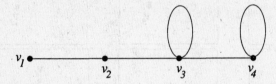

21.

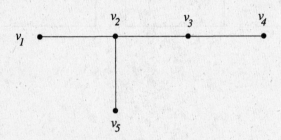

24. *b.* There are 7 nonempty subgraphs.

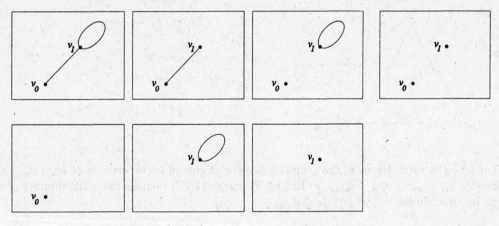

c. There are 17 nonempty subgraphs.

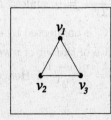 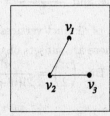 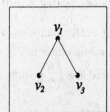 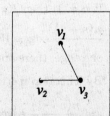

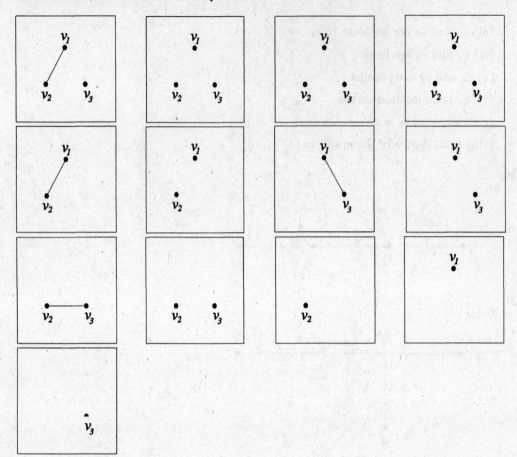

27. Yes. For example, the graph shown below satisfies this condition.

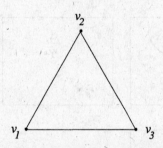

30. Let t be the total degree of the graph. Since the degree of each vertex is at least $d_{\min}$ and at most $d_{\max}$, $d_{\min} \cdot v \le t \le d_{\max} \cdot v$. But by Theorem 11.1.1, t equals twice the number of edges. So by substitution, $d_{\min} \cdot v \le 2e \le d_{\max} \cdot v$.

33. *b. Proof 1*: Suppose n is an integer with $n \ge 1$ and K_n is a complete graph on n vertices. If $n = 1$, then K_n has one vertex and 0 edges and $\dfrac{n(n-1)}{2} = \dfrac{1(1-1)}{2} = 0$, and so K_n has $\dfrac{n(n-1)}{2}$ edges. If $n \ge 2$, then since each pair of distinct vertices of K_n is connected by exactly one edge, there are as many edges in K_n as there are subsets of size two of the set of n vertices. By Theorem 6.4.1, there are $\dbinom{n}{2}$ such sets. But $\dbinom{n}{2} = \dfrac{n!}{2!(n-2)!} = \dfrac{n(n-1)}{2}$. Hence there are $\dfrac{n(n-1)}{2}$ edges in K_n.

Proof 2 (by mathematical induction): Let the property $P(n)$ be the sentence "the complete graph on n vertices, K_n, has $\dfrac{n(n-1)}{2}$ edges."

Show that the property is true for $n = 1$: *For $n = 1$ the property is true because the complete graph on one vertex, K_1, has 0 edges and $\dfrac{n(n-1)}{2} = \dfrac{1(1-1)}{2} = 0$.*

Show that for all integers $m \geq 1$, if the property is true for $n = m$ then it is true for $n = m + 1$: *Let m be an integer with $m \geq 1$, and suppose that K_m has $\dfrac{m(m-1)}{2}$ edges. [This is the inductive hypothesis.] We must show that K_{m+1} has $\dfrac{(m+1)((m+1)-1)}{2} = \dfrac{(m+1)m}{2}$ edges. Note that K_{m+1} vertices can be obtained from K_m vertices by adding one vertex, say v, and connecting v to each of the k other vertices. But by inductive hypothesis, K_m has $\dfrac{m(m-1)}{2}$ edges. Connecting v to each of the m other vertices adds another m edges. Hence the total number of edges of K_{m+1} is $\dfrac{m(m-1)}{2} + m = \dfrac{m(m-1)}{2} + \dfrac{2m}{2} = \dfrac{m^2 - m + 2m}{2} = \dfrac{m(m+1)}{2}$ [as was to be shown].*

Proof 3: Suppose n is an integer with $n \geq 1$ and K_n is a complete graph on n vertices. Because each vertex of K_n is connected by an edge to each of the other $n - 1$ vertices of K_n by exactly one edge, the degree of each vertex of K_n is $n - 1$. Thus the total degree of K_n equals the number of vertices times the degree of each vertex, or $n(n - 1)$. But by Theorem 11.1.1, the total degree of K_n equals twice the number e of edges of K_n, and so $n(n-1) = 2e$. Equivalently, $e = n(n - 1)/2$, [as was to be shown].

36. b. $K_{1,3}$

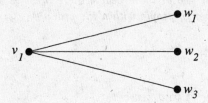

c. $K_{3,4}$

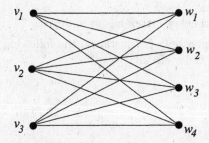

d. If $n \neq m$, the vertices of $K_{m,n}$ are divided into two groups: one of size m and the other of size n. Every vertex in the group of size m has degree n because each is connected to every vertex in the group of size n. So $K_{m,n}$ has n vertices of degree m. Similarly, every vertex in the group of size n has degree m because each is connected to every vertex in the group of size m. So $K_{m,n}$ has n vertices of degree m. Note that if $n = m$, then all $n + m = 2n$ vertices have the same degree, namely n.

e. The total degree of $K_{m,n}$ is $2mn$ because $K_{m,n}$ has m vertices of degree n (which contribute mn to its total degree) and n vertices of degree m (which contribute another mn to its total degree)

f. The number of edges of $K_{m,n} = mn$. The reason is that the total degree of $K_{m,n}$ is $2mn$, and so, by Theorem 11.1.1, $K_{m,n}$ has $2mn/2 = mn$ edges. Another way to reach this conclusion is to say that $K_{m,n}$ has n edges coming out of each of the group of m vertices (each leading to a vertex in the group of n vertices) for a total of mn edges. Equivalently, $K_{m,n}$ has m edges coming out of each of the group of n vertices (each leading to a vertex in the group of m vertices) for a total of mn edges.

39. *b.*

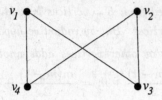

42. *The graph obtained by taking all the vertices and edges of G together with all the edges of G' is K_n. Therefore, by exercise 33b, the number of edges of G plus the number of edges of G' equals $n(n-1)/2$.*

ices has degree 0. Hence the supposition is false, and there is no simple graph with n vertices each of which has a different degree.

45. *Yes. Suppose that in a group of two or more people, each person is acquainted with a different number of people. Then the acquaintance graph representing the situation is a simple graph in which all the vertices have different degrees. But by exercise 44(c) such a graph does not exist. Hence the supposition is false, and so in a group of two or more people there must be at least two people who are acquainted with the same number of people within the group.*

Section 11.2

3. *b.* No, because $e_1 e_2$ could refer either to $v_1 e_1 v_2 e_2 v_1$ or to $v_2 e_1 v_1 e_2 v_2$.

6. *b.* $\{v_7, v_8\}, \{v_1, v_2\}, \{v_3, v_4\}$

   c. $\{v_2, v_3\}, \{v_6, v_7\}, \{v_7, v_8\}, \{v_9, v_{10}\}$

9. *b.* Yes, by Theorem 11.2.3 since G is connected and every vertex has even degree.

   c. Not necessarily. It is not specified that G is connected. For instance, the following graph satisfies the given conditions but does not have an Euler circuit:

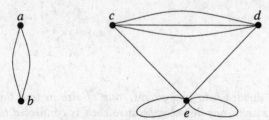

15. One Euler circuit is the following: *stuvwxyzrsuwyuzs*.

18. Yes. One Euler circuit is *ABDEACDA*.

21. One Euler path from u to w is $uv_1v_2v_3uv_0v_7v_6v_3v_4v_6wv_5v_4w$.

24. One Hamiltonian circuit is *balkjedcfihgb*.

27. Call the given graph G and suppose G has a Hamiltonian circuit. Then G has a subgraph H that satisfies conditions (1) – (4) of Proposition 11.2.6. Since the degree of B in G is five and every vertex in H has degree two, three edges incident on B must be removed from G to create H. Edge $\{B,C\}$ cannot be removed because doing so would result in vertex C having degree less than two in H. Similar reasoning shows that edges $\{B,E\}$, $\{B,F\}$, and $\{B,A\}$ cannot be removed either. It follows that the degree of B in H must be at least four, which contradicts the condition that every vertex in H has degree two in H. Hence no such subgraph H can exist, and so G does not have a Hamiltonian circuit.

30. One Hamiltonian circuit is $v_0v_1v_5v_4v_7v_6v_2v_3v_0$.

33. Other such graphs are those shown in exercises 17, 21, 23, 24, 29 and 30.

36. It is clear from the map that only a few routes have a chance of minimizing the distance. For instance, one must go to either Düsseldorf or Luxembourg just after leaving Brussels or just before returning to Brussels, and one must either travel from Berlin directly to Munich or the reverse. The possible minimizing routes are those shown below plus the same routes traveled in the reverse direction.

Route	Total Distance (in km)
Bru-Lux-Düss-Ber-Mun-Par-Bru	$219 + 224 + 564 + 585 + 832 + 308 = 2732$
Bru-Düss-Ber-Mun-Par-Lux-Bru	$223 + 564 + 585 + 832 + 375 + 219 = 2798$
Bru-Düss-Lux-Ber-Mun-Par-Bru	$223 + 224 + 764 + 585 + 832 + 308 = 2936$
Bru-Düss-Ber-Mun-Lux-Par-Bru	$223 + 564 + 585 + 517 + 375 + 308 = 2572$

The routes that minimize distance, therefore, are the bottom route shown in the table and that same route traveled in the reverse direction.

39. *Proof*: Suppose vertices v and w are part of a circuit in a graph G and one edge e is removed from the circuit. Without loss of generality, we may assume the v occurs before the w in the circuit, and we may denote the circuit by $v_0e_1v_1e_2\ldots e_{n-1}v_{n-1}e_nv_0$ with $v_i = v$, $v_j = w$, $i < j$, and $e_k = e$. If either $k \le i$ or $k > j$, then $v = v_ie_{i+1}v_{i+1}\ldots v_{j-1}e_jv_j = w$ is a path in G from v to w that does not include e. If $i < k \le j$, then $v = v_ie_iv_{i-1}e_{i-1}\ldots e_{j+1}v_j = w$ is a path in G from v to w that does not include e. These possibilities are illustrated by examples (1) and (2) in the diagram below. In either case there is a path in G from v to w that does not include e.

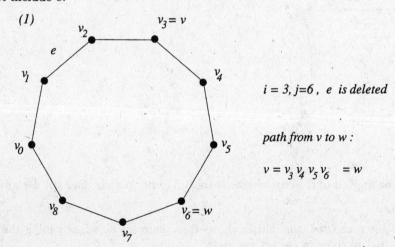

(1)

$i = 3, j=6$, e is deleted

path from v to w :

$v = v_3\ v_4\ v_5\ v_6\ = w$

(2)

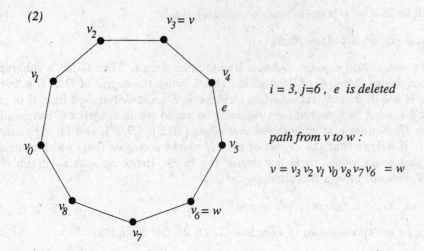

$i = 3, j = 6$, e is deleted

path from v to w :

$v = v_3 v_2 v_1 v_0 v_8 v_7 v_6 = w$

48. *a.* Let m and n be positive integers and let $K_{m,n}$ be a complete bipartite graph on (m, n) vertices. Since $K_{m,n}$ is connected, by Theorem 11.2.4 it has an Euler circuit if, and only if, every vertex has even degree. But $K_{m,n}$ has m vertices of degree n and n vertices of degree m. So $K_{m,n}$ has an Euler circuit if, and only if, both m and n are even.

b. Let m and n be positive integers, let $K_{m,n}$ be a complete bipartite graph on (m, n) vertices, and suppose $V_1 = \{v_1, v_2, \dots, v_m\}$ and $V_2 = \{w_1, w_2, \dots, w_n\}$ are the disjoint sets of vertices such that each vertex in V_1 is joined by an edge to each vertex in V_2 and no vertex within V_1 or V_2 is joined by an edge to any other vertex within the same set. If $m = n \geq 2$, then $K_{m,n}$ has the following Hamiltonian circuit: $v_1 w_1 v_2 w_2 \dots v_m w_m v_1$. If $K_{m,n}$ has a Hamiltonian circuit, then $m = n$ because the vertices in any Hamiltonian circuit must alternate between V_1 and V_2 (since no edges connect vertices within either set) and because no vertex, except the first and last, appears twice in a Hamiltonian circuit. If $m = n = 1$, then $K_{m,n}$ does not have a Hamiltonian circuit because $K_{1,1}$ contains just one edge joining two vertices. Therefore, $K_{m,n}$ has a Hamiltonian circuit if, and only if, $m = n \geq 2$.

Section 11.3

3. *b.*

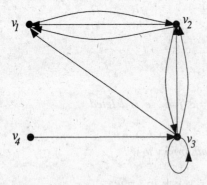

Any labels may be applied to the edges because the adjacency matrix does not determine edge labels.

6. *b.* The graph is not connected; the matrix shows that there are no edges joining the vertices from the set $\{v_1, v_2\}$ to those in the set $\{v_3, v_4\}$.

9. *b.*

$$\begin{bmatrix} 0 & 8 \\ -5 & 4 \end{bmatrix}$$

c.

$$\begin{bmatrix} -2 & 3 \\ 4 & 6 \end{bmatrix}$$

18. *Proof (by mathematical induction):* Let the property $P(n)$ be the sentence "A^n is symmetric."

Show that the property is true for $n = 1$: For $n = 1$ the property is true because by assumption A is a symmetric matrix.

Show that for all integers $k \geq 1$, if the property is true for $n = k$ then it is true for $n = k + 1$: Let k be an integer with $k \geq 1$, and suppose that A^k is symmetric. *[This is the inductive hypothesis.]* We must show that A^{k+1} is symmetric. Let $A^k = (b_{ij})$. Then for all $i, j = 1, 2, \ldots, m$, the ijth entry of A^{k+1} = the ijth entry of AA^k *[by definition of matrix power]* $= \Sigma_{r=1}^{m} a_{ir} b_{rj}$ *[by definition of matrix multiplication]* $= \Sigma_{r=1}^{m} a_{ri} b_{jr}$ *[because A is symmetric by hypothesis and A^k is symmetric by inductive hypothesis]* $= \Sigma_{r=1}^{m} b_{jr} a_{ri}$ *[because multiplication of real numbers is commutative]* = the jith entry of $A^k A$ *[by definition of matrix multiplication]* = the jith entry of AA^k *[by exercise 17]* = the jith entry of A^{k+1} *[by definition of matrix power]*. Therefore, A^{k+1} is symmetric *[as was to be shown]*.

21. *Proof (by mathematical induction):* Let the property $P(n)$ be the sentence "all the entries along the main diagonal of A^n are equal to each other and all the entries off the main diagonal are also equal to each other."

Show that the property is true for $n = 1$: For $n = 1$ the property is true because $A^1 = A$, which is the adjacency matrix for K_3, and all the entries along the main diagonal of A are 0 *[because K_3 has no loops]* and all the entries off the main diagonal are 1 *[because each pair of vertices is connected by exactly one edge]*.

Show that for all integers $m \geq 1$, if the property is true for $n = m$ then it is true for $n = m + 1$: Let m be an integer with $m \geq 1$, and suppose that all the entries along the main diagonal of A^m are equal to each other and all the entries off the main diagonal are also equal to each other. *[This is the inductive hypothesis.]* Then

$$A^m = \begin{bmatrix} b & c & c \\ c & b & c \\ c & c & b \end{bmatrix} \text{ for some integers } b \text{ and } c.$$

It follows that

$$A^{m+1} = AA^m = \begin{bmatrix} 0 & 1 & 1 \\ 1 & 0 & 1 \\ 1 & 1 & 0 \end{bmatrix} \begin{bmatrix} b & c & c \\ c & b & c \\ c & c & b \end{bmatrix} = \begin{bmatrix} 2c & b+c & b+c \\ b+c & 2c & b+c \\ b+c & b+c & 2c \end{bmatrix}$$

As can be seen, all the entries of A^{m+1} along the main diagonal are equal to each other and all the entries off the main diagonal are equal to each other. So the property is true for $n = m+1$.

Section 11.4

3. The graphs are isomorphic. One way to define to isomorphism is as follows.

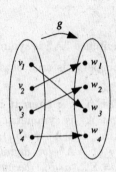
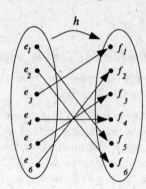

9. The graphs are isomorphic. One way to define to isomorphism is as follows.

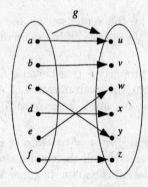

15. There is one such graph with 0 edges, one with 1 edge, and there are two with 2 edges, three with 3 edges, two with 4 edges, one with 5 edges, and one with 6 edges. These eleven graphs are shown below.

18. There are three such graphs in which all 3 edges are loops, five in which 2 edges are loops and 1 is not a loop, six in which 1 edge is a loop and 2 edges are not loops, and six in which none of the 3 edges is a loop. These twenty graphs are shown below.

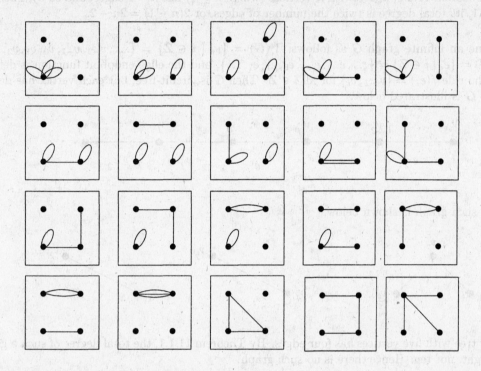

24. *Proof*: Suppose G and G' are isomorphic graphs and suppose G has a simple circuit C of length k, where k is a nonnegative integer. Let C be $v_0 e_1 v_1 e_2 \ldots e_k v_k (= v_0)$, and let C' be $g(v_0)h(e_1)g(v_1)h(e_2) \ldots h(e_k)g(v_k)(= g(v_0))$. By the same reasoning as in the solution to exercise 23, C' is a circuit of length k in G'. Suppose C' is not a simple circuit. Then C' has a repeated vertex, say $g(v_i) = g(v_j)$ for some $i, j = 0, 1, 2, \ldots, k - 1$ with $i \neq j$. But since g is a one-to-one correspondence this implies that $v_i = v_j$, which is impossible because C is a simple circuit. Hence the supposition is false, C' is a simple circuit, and therefore G' has a simple circuit of length k.

27. *Proof*: Suppose G and G' are isomorphic graphs and suppose G is connected. By definition of graph isomorphism, there are one-to-one correspondences $g: V(G) \to V(G')$ and $h: E(G) \to E(G')$ that preserve the edge-endpoint functions in the sense that for all v in $V(G)$ and e in $E(G)$, v is an endpoint of $e \Leftrightarrow g(v)$ is an endpoint of $h(e)$. Suppose w and x are any two vertices of G'. Then $u = g^{-1}(w)$ and $v = g^{-1}(x)$ are distinct vertices in G (because g is a one-to-one correspondence). Since G is connected, there is a walk in G connecting u and v. Say this walk is $ue_1 v_1 e_2 v_2 \ldots e_n v$. Because g and h preserve the edge-endpoint functions, $w = g(u)h(e_1)g(v_1)h(e_2)g(v_2) \ldots h(e_n)g(v) = x$ is a walk in G' connecting w and x.

30. Suppose that G and G' are isomorphic via one-to-one correspondences $g: V(G) \to V(G')$ and $h: E(G) \to E(G')$, where g and h preserve the edge-endpoint functions. Now w_6 has degree one in G', and so by the argument given in Example 11.4.4, w_6 must correspond to one of the vertices of degree one in G : either $g(v_1) = w_6$ or $g(v_6) = w_6$. Similarly, since w_5 has degree three in G', w_5 must correspond to one of the vertices of degree three in G : either $g(v_3) = w_5$ or $g(v_4) = w_5$. Because g and h preserve the edge-endpoint functions, edge f_6 with endpoints w_5 and w_6 must correspond to an edge in G with endpoints v_1 and v_3, or v_1 and v_4, or v_6 and v_3, or v_6 and v_4. But this contradicts the fact that none of these pairs of vertices are connected by edges in G. Hence the supposition is false, and G and G' are not isomorphic.

Section 11.5

3. By Theorem 11.5.2, a tree with n vertices (where $n \geq 1$) has $n - 1$ edges, and so by Theorem 11.1.1, its total degree is twice the number of edges, or $2(n - 1) = 2n - 2$.

6. Define an infinite graph G as follows: $V(G) = \{v_i \mid i \in \mathbf{Z}\} = \{\ldots, v_{-2}, v_{-1}, v_0, v_1, v_2, \ldots\}$, $E(G) = \{e_i \mid i \in \mathbf{Z}\} = \{\ldots, e_{-2}, e_{-1}, e_0, e_1, e_2, \ldots\}$, and the edge-endpoint function is defined by the rule $f(e_i) = \{v_{i-1}, v_i\}$ for all $i \in \mathbf{Z}$. Then G is circuit-free, but each vertex has degree two. G is illustrated below.

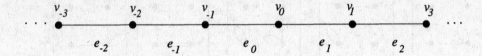

15. One such graph is shown below.

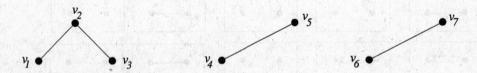

18. Any tree with five vertices has four edges. By Theorem 11.1.1, the total degree of such a graph is eight, not ten. Hence there is no such graph.

21. Any tree with ten vertices has nine edges. By Theorem 11.1.1, the total degree of such a tree is 18, not 24. Hence there is no such graph.

24. Yes. Given any two vertices u and w of G', then u and w are vertices of G neither equal to v. Since G is connected, there is a walk in G from u to w, and so by Lemma 11.2.1, there is a path in G from u to w. This path does not include edge e or vertex v because a path does not have a repeated edge, and e is the unique edge incident on v. *[If a path from u to w leads into v, then it must do so via e. But then it cannot emerge from v to continue on to w because no edge other than e is incident on v.]* Thus this path is a path in G'. It follows that any two vertices of G' are connected by a walk in G', and so G' is connected.

30. Such a tree must have 4 edges and, therefore, a total degree of 8. Since at least two vertices have degree 1 and no vertex has degree greater than 4, the possible degrees of the five vertices are as follows: 1,1,1,1,4; 1,1,1,2,3; and 1,1,2,2,2. The corresponding trees are shown below.

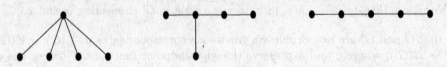

33. *a.* 3   *b.* 0   *c.* 5   *d.* v_{14}, v_{15}, v_{16} *e.* v_1 *f.* v_2 *g.* v_{17}, v_{18}, v_{19}

45. A full binary tree with k internal vertices has $2k + 1$ vertices in all. If $2k + 1 = 7$, then $k = 3$. Thus such a tree would have three internal and four terminal vertices. Two such trees are shown below.

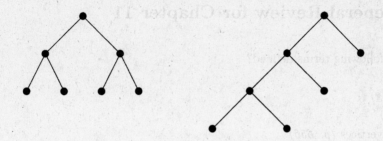

48. There is no tree with the given properties because a binary tree of height four has at most $2^4 = 16$ terminal vertices.

51. *b.* height $\geq \log_2 40 \cong 5.322$. Since the height of a tree is an integer, the height must be at least 6.

    c. height $\geq \log_2 60 \cong 5.907$. Since the height of a tree is an integer, the height must be at least 6.

Section 11.6

6. Minimum spanning tree:

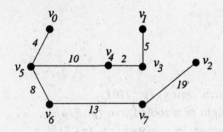

Order of adding the edges: $\{v_3, v_4\}$, $\{v_0, v_5\}$, $\{v_1, v_3\}$, $\{v_5, v_6\}$, $\{v_4, v_5\}$, $\{v_6, v_7\}$, $\{v_2, v_7\}$

12. *Proof*: Suppose T_1 and T_2 are spanning trees for a graph G with n vertices. By definition of spanning tree, both T_1 and T_2 contain all n vertices of G, and so by Theorem 11.5.2, both T_1 and T_2 have $n - 1$ edges.

18. *Proof*: Since T_2 is obtained from T_1 by removing e' and adding e, $w(T_2) = w(T_1) - w(e') + w(e)$. Now according to the proof of Theorem 11.6.3, $w(e') \geq w(e)$. Hence $w(e') - w(e) \geq 0$, and so $w(T_2) = w(T_1) - (w(e') - w(e)) \leq w(T_1)$.

24. The output will be a minimum spanning tree for the connected component of the graph that contains the starting vertex input to Prim's algorithm.

General Review for Chapter 11

Definitions: How are the following terms defined?

- graph *(p. 650)*
- directed graph *(p. 653)*
- simple graph *(p. 656)*
- complete graph on n vertices *(p. 656)*
- complete bipartite graph on (m, n) vertices *(p. 657)*
- subgraph *(p. 657)*
- degree of a vertex in a graph, total degree of a graph *(p. 658)*
- walk, path, simple path, closed walk, circuit, simple circuit *(p. 667)*
- trivial circuit, nontrivial circuit *(p. 669)*
- connected vertices, connected graph *(p. 669)*
- connected component of a graph *(p. 670)*
- Euler circuit in a graph *(p. 671)*
- Euler path in a graph *(p. 675)*
- Hamiltonian circuit in a graph *(p. 677)*
- adjacency matrix of a directed (or undirected) graph *(pp. 685-6)*
- symmetric matrix *(p. 687)*
- isomorphic graphs *(p. 698)*
- isomorphic invariant for graphs *(p. 701)*
- circuit-free graph *(p. 705)*
- tree *(p. 705)*
- terminal vertex (or leaf), internal vertex (or branch vertex) *(p. 710)*
- rooted tree, level of a vertex in a rooted tree, height of a rooted tree *(p. 715)*
- parents, children, siblings, descendants, and ancestors in a rooted tree *(p. 715)*
- binary tree, full binary tree *(p. 716)*
- spanning tree *(p. 724)*
- weighted graph, minimum spanning tree *(p. 725)*

Graphs

- What does the handshake theorem say? In other words, how is the total degree of a graph related to the number of edges of the graph? *(p. 659)*
- How can you use the handshake theorem to determine whether graphs with specified properties exist? *(pp. 660, 662)*
- If an edge is removed from a nontrivial circuit in a graph, does the graph remain connected? *(p. 670)*
- A graph has an Euler circuit if, and only if, it satisfies what two conditions? *(p. 675)*
- A graph has a Hamiltonian circuit if, and only if, it satisfies what four conditions? *(p. 678)*
- What is the traveling salesman problem? *(p. 679)*
- How do you find the adjacency matrix of a directed (or undirected) graph? How do you find the graph that corresponds to a given adjacency matrix? *(pp. 685-6)*
- How can you determine the connected components of a graph by examining the adjacency matrix of the graph? *(p. 688)*
- How do you multiply two matrices? *(p. 689)*
- How do you use matrix multiplication to compute the number of walks from one vertex to another in a graph? *(p. 694)*

- How do you show that two graphs are isomorphic? *(p. 698)*
- What are some invariants for graph isomorphisms? *(p. 701)*

Trees

- If a tree has at least two vertices, how many vertices of degree 1 does it have? *(p. 709)*
- If a tree has n vertices, how many edges does it have? *(p. 710)*
- If a connected graph has n vertices, what additional property guarantees that it will be a tree? *(p. 714)*
- Given a full binary tree, what is the relation among the number of its internal vertices, terminal vertices, and total number of vertices? *(p. 717)*
- Given a binary tree, what is the relation between the number of its terminal vertices and its height? *(p. 718)*
- How does Kruskal's algorithm work? *(p. 726)*
- How do you know that Kruskal's algorithm produces a minimum spanning tree? *(p. 727)*
- How does Prim's algorithm work? *(p. 729)*
- How do you know that Prim's algorithm produces a minimum spanning tree? *(p. 730)*

Test Your Understanding: Chapter 11

Test yourself by filling in the blanks.

1. A graph consists of two finite sets: _____ and _____, where each edge is associated with a set consisting of _____.

2. A loop in a graph is _____.

3. Two distinct edges in a graph are parallel if, and only if, _____.

4. An edge is said to _____ its endpoints.

5. Two vertices are called adjacent if, and only if, _____.

6. An edge is incident on _____.

7. Two edges incident on the same endpoint are _____.

8. A vertex on which no edges are incident is _____.

9. A graph with no vertices is _____.

10. In a directed graph, each edge is associated with _____.

11. A simple graph is _____.

12. A complete graph on n vertices is a _____.

13. A complete bipartite graph on (m, n) vertices is a simple graph whose vertices can be divided into two distinct sets V_1 and V_2 in such a way that (1) each of the m vertices in V_1 is _____ to each of the n vertices in V_2, no vertex in V_1 is connected to _____, and no vertex in V_2 is connected to _____.

14. A graph H is a subgraph of a graph G if, and only if, (1) _____, (2) _____, and (3) _____.

15. The degree of a vertex in a graph is _____.

16. The total degree of a graph is ____.

17. The handshake theorem says that the total degree of a graph is ____.

18. In any graph the number of vertices of odd degree is ____.

19. Let G be a graph and let v and w be vertices in G.
    (a) A walk from v to w is ____.
    (b) A path from v to w is ____.
    (c) A simple path from v to w is ____.
    (d) A closed walk is ____.
    (e) A circuit is ____.
    (f) A simple circuit is ____.
    (g) A trivial circuit is ____.
    (h) Vertices v and w are connected if, and only if, ____.

20. A graph is connected if, and only if, ____.

21. Removing an edge from a nontrivial circuit in a graph does not ____.

22. An Euler circuit in a graph is ____.

23. A graph has an Euler circuit if, and only if, ____.

24. Given vertices v and w in a graph, there is an Euler path from v to w if, and only if, ____.

25. A Hamiltonian circuit in a graph is ____.

26. If a graph G has a nontrivial Hamiltonian circuit, then G has a subgraph H with the following properties: ____, ____, ____, and ____.

27. A traveling salesman problem involves finding a ____ that minimizes the total distance traveled for a graph in which each edge is marked with a distance.

28. In an adjacency matrix for a directed graph, the entry in the ith row and jth column is ____.

29. In an adjacency matrix for a (undirected) graph, the entry in the ith row and jth column is ____.

30. An $n \times n$ square matrix is called symmetric if, and only if, for all integers i and j from 1 to n, the entry in row ____ and column ____ equals the entry in row ____ and column ____.

31. The ijth entry in the product of two matrices $\mathbf{A}$ and $\mathbf{B}$ is obtained by multiplying row ____ of $\mathbf{A}$ by row ____ of $\mathbf{B}$.

32. In an $n \times n$ identity matrix the entries along the diagonal are all ____ and the off-diagonal entries are all ____.

33. If G is a graph with vertices $v_1, v_2, \ldots, v_m$ and $\mathbf{A}$ is the adjacency matrix of G, for each positive integer n and for all integers i and j with $i, j = 1, 2, \ldots, m$, the ijth entry of $\mathbf{A}^n = $ ____.

34. If G and G' are graphs, then G is isomorphic to G' if, and only if, there exist a one-to-one correspondence g from the vertex set of G to the vertex set of G' and a one-to-one correspondence h from the edge set of G to the edge set of G' such that for all vertices v and edges e in G, v is an endpoint of e if, and only if, ____.

35. A property P is an isomorphic invariant for graphs if, and only if, given any graphs G and G', if G has property P and G' is isomorphic to G then ____.

36. Some invariant properties for graph isomorphisms are ____, ____, ____, ____, ____, ____, ____, ____, ____, and ____.

37. A circuit-free graph is a graph with ____.

38. A forest is a graph that is ____, and a tree is a graph that is ____.

39. A trivial tree is a graph that consists ____, and an empty tree is a tree that ____.

40. Any tree with at least two vertices has at least one vertex of degree ____.

41. If a tree T has at least two vertices, then a terminal vertex (or leaf) in T is a vertex of degree ____ and an internal vertex (or branch vertex) in T is a vertex of degree ____.

42. For any positive integer n, any tree with n vertices has ____.

43. For any positive integer n, if G is a connected graph with n vertices and $n - 1$ edges then ____.

44. A rooted tree is a tree in which ____. The level of a vertex in a rooted tree is ____. The height of a rooted tree is ____.

45. A binary tree is a rooted tree in which ____.

46. A full binary tree is a rooted tree in which ____.

47. If k is a positive integer and T is a full binary tree with k internal vertices, then T has a total of ____ vertices and has ____ terminal vertices.

48. If T is a binary tree that has t terminal vertices and height h, then t and h are related by the inequality ____.

49. A spanning tree for a graph G is ____.

50. A weighted graph is a graph for which ____, and the total weight of the graph is ____.

51. A minimum spanning tree for a connected weighted graph is ____.

52. In Kruskal's algorithm, the edges of a connected, weighted graph are examined one by one in order of ____.

53. In Prim's algorithm, a minimum spanning tree is built by expanding outward ____.

Answers

1. a finite set of vertices, a finite set of edges, either one or two vertices called its endpoints
2. an edge with a single endpoint
3. they have the same set of endpoints
4. connect
5. they are connected by an edge
6. each of its endpoints
7. adjacent
8. isolated
9. empty
10. an ordered pair of vertices called its endpoints
11. a graph with no loops or parallel edges
12. simple graph with n vertices whose set of edges contains exactly one edge for each pair of vertices

13. connected by an edge, any other vertex in V_1, any other vertex in V_2

14. every vertex in H is also a vertex in G, every edge in H is also a vertex in G, every edge in H has the same endpoints as it has in G

15. the number of edges that are incident on the vertex, with an edge that is a loop counted twice

16. the sum of the degrees of all the vertices of the graph

17. equal to twice the number of edges of the graph

18. an even number

19.
    (a) a finite alternating sequence of adjacent vertices and edges of G
    (b) a walk that does not contain a repeated edge
    (c) a path that does not contain a repeated vertex
    (d) a walk that starts and ends at the same vertex
    (e) a closed walk that does not contain a repeated edge
    (f) a circuit that does not have any repeated vertex other than the first and the last
    (g) a walk consisting of a single vertex and no edge
    (h) there is a walk from v to w

20. given any two vertices in the graph there is a walk from one to the other

21. disconnect the graph

22. a circuit that contains every vertex and every edge of the graph

23. the graph is connected and every vertex has even degree

24. the graph is connected, v and w have odd degree, and all other vertices have even degree

25. a simple circuit that includes every vertex of the graph

26. H contains every vertex of G; H is connected; H has the same number of edges as vertices; every vertex of H has degree 2

27. Hamiltonian circuit

28. the number of arrows from v_i (the ith vertex) to v_j (the jth vertex)

29. the number of edges connecting v_i (the ith vertex) and v_j (the jth vertex)

30. i; j; j; i

31. i; j

32. 1; 0

33. the number of walks of length n from v_i to v_j

34. $g(v)$ is an endpoint of $h(e)$

35. G' has property P

36. has n vertices; has m edges; has a vertex of degree k; has m vertices of degree k; has a circuit of length k; has a simple circuit of length k; has m simple circuits of length k; is connected; has an Euler circuit; has a Hamiltonian circuit

37. no nontrivial circuits

38. circuit-free; connected and circuit-free

39. of a single vertex and no edges; has no vertices or edges

40. 1

41. 1; at least 2

42. $n - 1$ edges

43. G is a tree

44. one vertex is distinguished from the others and is called the root
the number of edges along the unique path between it and the root
the maximum level of any vertex of the tree

45. every parent has at most two children

46. every parent has exactly two children

47. $2k + 1$; $k + 1$

48. $t \leq 2^h$, or, equivalently, $\log_2 t \leq h$

49. a subgraph of G that contains every vertex of G and is a tree

50. each edge has an associated real number weight
    the sum of the weights of all the edges of the graph

51. a spanning tree that has the least possible total weight compared to all other spanning trees
    for the graph

52. weight, starting with an edge of least weight

53. in a sequence of adjacent edges starting from some vertex

Chapter 12: Regular Expressions and Finite-State Automata

This chapter opens with some historical background about the connections between computers and formal languages. Section 12.1 focuses on regular expressions and emphasizes their utility for pattern matching, whether for compilers or for general text processing.

Section 12.2 introduces the concept of finite-state automaton. In one sense, it is a natural sequel to the discussions of digital logic circuits in Section 1.4 and Boolean functions in Section 7.1, with the next-state function of an automaton governing the operation of sequential circuit in much the same way that a Boolean function governs the operation of a combinatorial circuit. The section also provides practice in finding a finite-state automaton that corresponds to a regular expression and shows how to write a program to implement a finite-state automaton. Both abilities are useful for computer programming. The section ends with a statement and partial proof of Kleene's theorem, which describes the exact nature of the relationship between finite-state automata and regular languages.

The equivalence and simplification of finite-state automata, discussed in Section 12.3, provides an additional application for the concept of equivalence relation, introduced in Section 10.3. Note the parallel between the simplification of digital logic circuits discussed in Section 1.4 and the simplification of finite-state automata developed in this section. Both kinds of simplification have obvious practical use.

Section 12.1

3. $L = \{11*, 11/, 12*, 12/, 21*, 21/, 22*, 22/\}$

   $11* = 1*1 = 1,\ 11/ = 1/1 = 1,\ 12* = 1*2 = 2,\ 12/ = 1/2 = 0.5,\ 21* = 2*1 = 2,$
   $21/ = 2/1 = 2,\ 22* = 2*2 = 4,\ 22/ = 2/2 = 1$

6. $L_1 L_2$ is the set of strings of 0's and 1's that both start and end with a 0.

   $L_1 \cup L_2$ is the set of strings of 0's and 1's that start with a 0 or end with a 0 (or both).

   $(L_1 \cup L_2)^*$ is the set of strings of 0's and 1's that start with a 0 or end with a 0 (or both) or that contain 00.

9. $(((x \mid (y(z^*)))^*)((yx) \mid (((yz)^*)z))$

12. $xy(x^*y)^* \mid (yx \mid y)y^*$

15. $L((a \mid b)c) = L(a \mid b)L(c) = (L(a) \cup L(b))L(c) = (\{a\} \cup \{b\})\{c\} = \{a, b\})\{c\} = \{ac, bc\}$

18. $x, yxxy, xx, xyxxy, xyxxyyxxy, \ldots$

21. The language consists of the set of all strings of x's and y's that start with xy or yy followed by any string of x's and y's.

24. The string 120 does not belong to the language defined by $(01^*2)^*$ because it does not start with 0. However, 01202 does belong to the language because 012 and 02 are both defined by 01^*2 and the language is closed under concatenation.

27. $x \mid y^* \mid y^*(xyy^*)(\epsilon \mid x)$

30. Note that for any regular expression x, $(x^*)^*$ defines the set of all strings obtained by concatenating a finite number of a finite number of concatenations of copies of x. But any such string can equally well be obtained simply by concatenating a finite number of copies of x, and thus $(x^*)^* = x^*$. Hence the given languages are the same: $L((rs)^*) = L(((rs)^*)^*)$.

33. $[a - z]\{3\}[a - z]^*ly$

36. $[^\char`\^AEIOU][A-Z]^*[A|E|I|O|U]\{2\}[A-Z]^*$

Section 12.2

3. *a.* U_0, U_1, U_2, U_3 *b.* a, b *c.* U_0 *d.* U_3

   e.

		input	
		a	b
$\rightarrow$	U_0	U_2	U_1
state	U_1	U_2	U_3
	U_2	U_2	U_2
◎	U_3	U_3	U_3

6. *a.* s_0, s_1, s_2, s_3 *b.* $0, 1$ *c.* s_0 *d.* s_0

   e.

			input	
			0	1
$\rightarrow$	◎	s_0	s_0	s_1
state		s_1	s_1	s_2
		s_2	s_2	s_3
		s_3	s_3	s_0

9. *a.* s_0, s_1, s_2, s_3 *b.* $0, 1$ *c.* s_0 *d.* s_1

   e.

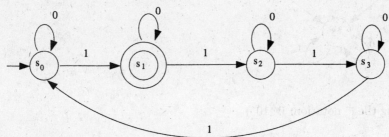

21. *a.*

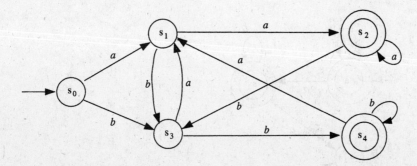

   b. $(a|b)^*(aa|bb)$

24. *a.*

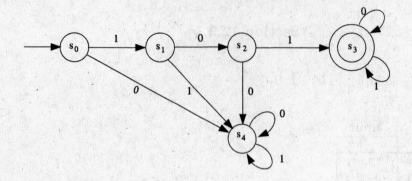

  b. 101(0|1)*

27. *a.*

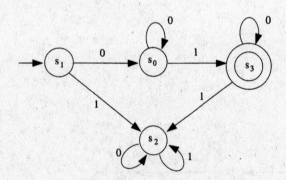

  b. 00*10* (or using the + notation: 0⁺10*)

30.

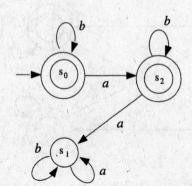

48. Let d represent the character class $[0-9]$.

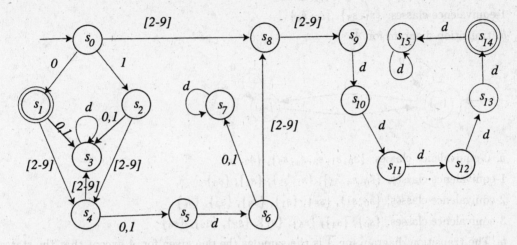

51. *Proof (by contradiction):* Suppose there were a finite-state automaton A that accepts L. Consider all strings of the form a^i for some integer $i \geq 0$. Since the set of all such strings is infinite and the number of states of A is finite, by the pigeonhole principle at least two of these strings, say a^p and a^q with $p < q$, must send A to the same state, say s, when input to A starting in its initial state. (The strings of the given form are the pigeons, the states are the pigeonholes, and each string is associated with the state to which A goes when the string is input to A starting in its initial state.) Because A accepts L, A accepts $a^q b^q$ but not $a^p b^q$. But since $a^q b^q$ is accepted by A, inputting b^q to A when it is in state s (after input of a^q) sends A to an accepting state. Because A also goes to state s after input of a^p, inputting b^q to A after inputting a^p also sends A to an accepting state. Thus $a^p b^q$ is accepted by A and yet it is not accepted by A, which is a contradiction. Hence the supposition is false: there is no finite-state automaton that accepts L.

54. *a. Proof:* Suppose A is a finite-state automaton with input alphabet $\sum$, and suppose $L(A)$ is the language accepted by A. Define a new automaton A' as follows: Both the states and the input symbols of A' are the same as the states and input symbols of A. The only difference between A and A' is that each accepting state of A is a non-accepting state of A', and each non-accepting state of A is an accepting state of A'. It follows that each string in $\sum^*$ that is accepted by A is not accepted by A', and each string in $\sum^*$ that is not accepted by A is accepted by A'. Thus $L(A') = (L(A))^c$.

b. Proof: Let A_1 and A_2 be finite-state automata, and let $L(A_1)$ and $L(A_2)$ be the languages accepted by A_1 and A_2, respectively. By part (a), there exist automata A_1' and A_2' such that $L(A_1') = (L(A_1))^c$ and $L(A_2') = (L(A_2))^c$. Hence, by Kleene's theorem (part 1), there are regular expressions r_1 and r_2 that define $(L(A_1))^c$ and $(L(A_2))^c$, respectively. So we may write $(L(A_1))^c = L(r_1)$ and $(L(A_2))^c = L(r_2)$. Now by definition of regular expression, $r_1 \mid r_2$ is a regular expression, and, by definition of the language defined by a regular expression, $L(r_1 \mid r_2) = L(r_1) \cup L(r_2)$. Thus, by substitution and De Morgan's law, $L(r_1 \mid r_2) = (L(A_1))^c \cup (L(A_2))^c = (L(A_1) \cap L(A_2))^c$, and so, by Kleene's theorem (part(2)), there is a finite-state automaton, say A, that accepts $(L(A_1) \cap L(A_2))^c$. It follows from part (a) that there is a finite-state automaton, A', that accepts $((L(A_1) \cap L(A_2))^c)^c$. But, by the double complement law for sets, $((L(A_1) \cap L(A_2))^c)^c = L(A_1) \cap L(A_2)$. So there is a finite-state automaton, A', that accepts $L(A_1) \cap L(A_2)$, and hence, by Kleene's theorem and the definition of regular language, $L(A_1) \cap L(A_2)$ is a regular language.

Section 12.3

3. *a.* 0-equivalence classes: $\{s_1, s_3\}, \{s_0, s_2\}$

   1-equivalence classes: $\{s_1, s_3\}, \{s_0, s_2\}$

   b. transition diagram for $\bar{A}$:

   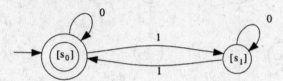

6. *a.* 0-equivalence classes: $\{s_0, s_1, s_3, s_4, s_5\}, \{s_2, s_6\}$

   1-equivalence classes: $\{s_0, s_4, s_5\}, \{s_1, s_3\}, \{s_2\}, \{s_6\}$

   2-equivalence classes: $\{s_0, s_4\}, \{s_5\}, \{s_1\}, \{s_3\}, \{s_2\}, \{s_6\}$

   3-equivalence classes: $\{s_0\}, \{s_4\}, \{s_5\}, \{s_1\}, \{s_3\}, \{s_2\}, \{s_6\}$

   b. The transition diagram for $\overline{A}$ is the same as the one given for A except that the states are denoted $[s_0], [s_1], [s_2], [s_3], [s_4], [s_5], [s_6]$.

15. *Proof*: Suppose k is an integer such that $k \geq 1$ and C_k is a k-equivalence class. We must show that there is a $k-1$ equivalence class, C_{k-1}, such that $C_k \subseteq C_{k-1}$. By property (12.3.3), the $(k-1)$-equivalence classes partition the set of all states of A in to a union of mutually disjoint subsets. Let s be any state in C_k. Then s is in *some* $(k-1)$-equivalence class; call it C_{k-1}. Let t be any other state in C_k. *[We will show that $t \in C_{k-1}$ also.]* Then $t\, R_k\, s$, and so for all input strings of length k, $N^*(t, w)$ is an accepting state $\Leftrightarrow N^*(s, w)$ is an accepting state. Since $k - 1 < k$, it follows that for all input strings of length $k - 1$, $N^*(t, w)$ is an accepting state $\Leftrightarrow N^*(s, w)$ is an accepting state. Consequently, $t\, R_{k-1}\, s$, and so t and s are in the same $(k-1)$-equivalence class. But $s \in C_{k-1}$. Hence $t \in C_{k-1}$ also. We, therefore, conclude that $C_k \subseteq C_{k-1}$.

18. *Proof*: Suppose A is an automaton and C is a $*$-equivalence class of states of A. By Theorem 12.3.2, for some integer $K \geq 0$, C is a K-equivalence class of A. Suppose C contains both an accepting state s and a nonaccepting state t of A. Since both s and t are in the same K-equivalence class, s is K-equivalent to t (by exercise 34 of Section 10.3), and so by exercise 17, s is 0-equivalent to t. But this is impossible because there are only two 0-equivalence classes, the set of all accepting states and the set of all nonaccepting states, and these two sets are disjoint. Hence the supposition that C contains both an accepting and a nonaccepting state is false: C consists entirely of accepting states or entirely of nonaccepting states.

General Review for Chapter 12

Definitions: How are the following terms defined?

- alphabet, string over an alphabet, formal language over an alphabet *(p. 736)*
- Σ^n, Σ^* (the Kleene closure of Σ), and Σ^+ (the positive closure of Σ), where Σ is an alphabet *(p. 736)*
- concatenation of x and y, where x and y are strings *(p. 738)*
- concatenation of L and L', where L and L' are languages *(p. 738)*
- union of L and L', where L and L' are languages *(p. 738)*
- Kleene closure of L , where L is a language *(p. 738)*
- regular expression over an alphabet *(p. 738)*
- language defined by a regular expression *(p. 739)*
- finite-state automaton *(p. 748)*
- language accepted by a finite-state automaton *(p. 750)*
- eventual-state function for a finite-state automaton *(p. 751)*
- regular language *(p. 759)*
- *-equivalence of states in a finite-state automaton *(p. 764)*
- k-equivalence of states in a finite-state automaton *(p. 765)*
- quotient automaton *(p. 769)*
- equivalent automata *(p. 771)*

Regular Expressions

- What is the order of precedence for the operations in a regular expression? *(p. 737)*
- How do you find the language defined by a regular expression? *(p. 740)*
- Given a language, how do you find a regular expression that defines the language? *(p. 741)*
- What are some practical uses of regular expressions? *(p. 742)*

Finite-State Automata

- How do you construct an annotated next-state table for a finite-state automaton given the transition diagram for the automaton? *(p. 748)*
- How do you construct a transition diagram for a finite-state automaton given its next-state table? *(p. 749)*
- How do you find the state to which a finite-state automaton goes if the characters of a string are input to it? *(p. 750)*
- How do you find the language accepted by a finite-state automaton? *(p. 750)*
- Given a simple formal language, how do you construct a finite-state automaton to accept the language? *(p. 752)*
- How can you use software to simulate the action of a finite-state automaton? *(p. 754)*
- What do the two parts of Kleene's theorem say about the relation between the language accepted by a finite-state automaton and the language defined by a regular expression? *(pp. 756, 758)*
- How can the pigeonhole principle be used to show that a language is not regular? *(p. 759)*
- How do you find the k-equivalence classes for a finite-state automaton? *(p. 766)*
- How do you find the *-equivalence classes for a finite-state automaton? *(p. 767)*
- How do you construct the quotient automaton for a finite-state automaton? *(p. 769)*
- What is the relation between the language accepted by a finite-state automaton and the language accepted by the corresponding quotient automaton? *(p. 769)*

Test Your Understanding: Chapter 12

Test yourself by filling in the blanks.

1. If x and y are strings, the concatenation of x and y is ____.

2. If L and L' are languages, the concatenation of L and L' is ____.

3. If L and L' are languages, the union of L and L' is ____.

4. If L is a language, the Kleene closure of L is ____.

5. The set of regular expressions over a finite alphabet Σ is defined recursively. The BASE for the definition is the statement that ____. The RECURSION for the definition specifies that if r and s are any regular expressions in the set, then the following are also regular expressions in the set: ____, ____, and ____.

6. The function that associates a language to each regular expression over an alphabet Σ is defined recursively. The BASE for the definition is the statement that $L(\emptyset) =$ ____, $L(\epsilon) =$ ____, and $L(a) =$ ____ for every $a \in \Sigma$. The RECURSION for the definition specifies that if $L(r)$ and $L(r')$ are the languages defined by the regular expressions r and r' over Σ, then $L(rr') =$ ____, $L(r \mid r') =$ ____, and $L(r^*) =$ ____.

7. The notation $[A - C\ x - z]$ is an example of a ____ and denotes the regular expression ____.

8. Use of a single dot in a regular expression stands for ____.

9. The symbol $\wedge$, placed at the beginning of a character class, indicates ____.

10. The symbol $+$ following a regular expression r means that ____.

11. If r is a regular expression, the notation $r?$ denotes ____.

12. If r is a regular expression, the notation $r\{n\}$ means that ____ and the notation $r\{m, n\}$ means that ____.

13. The five objects that make up a finite-state automaton are ____, ____, ____, ____, and ____.

14. The next-state table for an automaton shows the values of ____.

15. In the annotated next-state table, the initial state is indicated with an ____ and the accepting states are marked by ____.

16. A string w consisting of input symbols is accepted by a finite-state automaton A if, and only if, ____.

17. The language accepted by a finite-state automaton A is ____.

18. If N is the next-state function for a finite-state automaton A, the eventual-state function N^* is defined as follows: for each state s of A and for each string w that consists of input symbols of A, $N^*(s, w) =$ ____.

19. One part of Kleene's theorem says that given any language that is accepted by a finite-state automaton, there is ____.

20. The second part of Kleene's theorem says that given any language defined by a regular expression, there is ____.

21. A regular language is ____.

22. Given the language consisting of all strings of the form $a^k b^k$, where k is a positive integer, the pigeonhole principle can be used to show that the language is ____.

23. Given a finite-state automaton A with eventual-state function N^* and given any states s and t in A, we say that s and t are $*$-equivalent if, and only if, ____.

24. Given a finite-state automaton A with eventual-state function N^* and given any states s and t in A, we say that s and t are k-equivalent if, and only if, ____.

25. Given states s and t in a finite-state automaton A, s is 0-equivalent to t if, and only if, either both s and t are ____ or both are ____. Moreover, for every integer $k \geq 1$, s is k-equivalent to t if, and only if, (1) s and t are $(k-1)$-equivalent and (2) ____.

26. If A is a finite-state automaton, then for some integer $K \geq 0$, the set of K-equivalence classes of states of A equals the set of ____-equivalence classes of A, and for all such K these are both equal to the set of ____.

27. Given a finite-state automaton A, the set of states of the quotient automaton $\bar{A}$ is ____.

Answers

1. the string obtained by juxtaposing the characters of x and y
2. $\{xy \mid x \in L \text{ and } y \in L'\}$
3. $\{s \mid s \in L \text{ or } s \in L'\}$
4. $\{t \mid t \text{ is a concatenation of any finite number of strings in } L\}$
5. $\emptyset$, ϵ, and each individual symbol in Σ are regular expressions over Σ; (rs); $(r \mid s)$; (r^*)
6. $\emptyset$; $\{\epsilon\}$; $\{a\}$; $L(r)L(r')$; $L(r) \cup L(r')$; $(L(r))^*$
7. character class; $(A \mid B \mid C \mid x \mid y \mid z)$
8. an arbitrary character
9. a character of the same type as those in the range of the class, but not any of the characters following the $^\wedge$, is to occur at that point in the string
10. the string contains at least one occurrence of r
11. $(\epsilon \mid r)$
12. r can be concatenated with itself n times; r can be concatenated with itself from m through n times
13. a finite set of states; a finite set of input symbols; a designated initial state; a designated set of accepting states; a next-state function that associates a "next-state" with each state and input symbol of the automaton
14. the next-state function for each state and input symbol of the automaton
15. arrow; double circles
16. when the symbols in the string are input to the automaton in sequence from left to right, starting from the initial state, the automaton ends up in an accepting state
17. the set of strings that are accepted by A
18. the state to which A goes if it is in state s and the characters of w are input to it in sequence
19. a regular expression that defines the same language
20. a finite-state automaton that accepts the same language
21. a language defined by a regular expression
22. not regular
23. for all input strings w, either $N^*(s, w)$ and $N^*(t, w)$ are both accepting states or both are nonaccepting states
24. for all input strings w of length less than or equal to k, either $N^*(s, w)$ and $N^*(t, w)$ are both accepting states or both are nonaccepting states
25. accepting states, nonaccepting states; for any input symbol m, $N(s, m)$ and $N(t, m)$ are also $(k-1)$-equivalent
26. $(K+1)$; $*$-equivalence classes of states of A
27. the set of $*$-equivalence classes of states of A